Synchrotron Radiation and Magnetism: Light Sources, Techniques and Applications

Synchrotron Radiation and Magnetism: Light Sources, Techniques and Applications

Susan Darby

NY RESEARCH
P R E S S

New York

Published by NY Research Press
118-35 Queens Blvd., Suite 400,
Forest Hills, NY 11375, USA
www.nyresearchpress.com

Synchrotron Radiation and Magnetism: Light Sources, Techniques and Applications
Susan Darby

International Standard Book Number: 978-1-64725-419-3 (Hardback)

Cataloging-in-Publication Data

Synchrotron radiation and magnetism : light sources, techniques and applications / Susan Darby.
p. cm.
Includes bibliographical references and index.
ISBN 978-1-64725-419-3
1. Synchrotron radiation. 2. Magnetism. 3. Physics.
4. Light sources. I. Darby, Susan.
QC793.5.E627 S96 2023
539.735--dc23

Contents

Preface

This book was inspired by the evolution of our times; to answer the curiosity of inquisitive minds. Many developments have occurred across the globe in the recent past which has transformed the progress in the field.

Synchrotron radiation (SR) refers to the electromagnetic radiation produced when relativistic charged particles experience an acceleration perpendicular to their velocity. It is created artificially in particle accelerators or naturally by fast electrons moving through magnetic fields. An SR source is a source that emits electromagnetic radiation (EM) through a storage ring or bending magnets for technical and scientific purposes. The techniques used in SR include imaging and microscopy, X-ray diffraction and scattering, and spectroscopy. SR has numerous applications in medical imaging, particularly, in mammography, monochromatic computed tomography, intravenous coronary angiography and bronchography. It has shown to be an extremely powerful instrument for the study of magnetic materials and phenomena due to its unique properties including adaptable time structure, wide energy spectrum, and variable light polarization. This book aims to understand the various light sources, techniques and applications of synchrotron radiation and magnetism. It is appropriate for students seeking detailed information in this area of study as well as for experts.

This book was developed from a mere concept to drafts to chapters and finally compiled together as a complete text to benefit the readers across all nations. To ensure the quality of the content we instilled two significant steps in our procedure. The first was to appoint an editorial team that would verify the data and statistics provided in the book and also select the most appropriate and valuable contributions from the plentiful contributions we received from authors worldwide. The next step was to appoint an expert of the topic as the Editor-in-Chief, who would head the project and finally make the necessary amendments and modifications to make the text reader-friendly. I was then commissioned to examine all the material to present the topics in the most comprehensible and productive format.

I would like to take this opportunity to thank all the contributing authors who were supportive enough to contribute their time and knowledge to this project. I also wish to convey my regards to my family who have been extremely supportive during the entire project.

Susan Darby

1

An Introductory Overview of Large-Scale Facilities and X-Ray Sources

Philip R. Willmott

Abstract High-brilliance X-ray sources are powerful probes to investigate the properties of matter down to the sub-angstrom scale and on time scales that can extend below a femtosecond. In this chapter, an introductory overview of the physics behind storage ring-based synchrotrons and linear accelerator-based X-ray free-electron lasers is presented, while the properties of the radiation they produce are explained.

1.1 Introduction

Since their discovery by Wilhelm Röntgen in the last decade of the nineteenth century, X-rays have played a central role in all branches of the natural sciences and medicine. The primary reasons for this are threefold. Firstly, 'hard' X-rays (that is, those with photon energies in excess of a few keV and up to approximately 50 keV, see Fig. 1.1), have interaction strengths that, on the one hand, are small enough to allow them to penetrate deeply into solid matter, while, on the other, are sufficiently large that these interactions are easily observable. Secondly, X-rays have wavelengths λ on the nanometre to angstrom scale, meaning that, according to the Abbe limit, they can be used to image objects composed of features down to sizes comparable to λ. Finally, the binding energies of electrons, from weakly bound valence electrons to very strongly bound core electrons in heavy elements, lie in the ultraviolet to hard X-ray regime, allowing detailed studies of these bonds through spectroscopy, thereby providing insights into chemistry, electronic structure and magnetic properties.

Modern scientific disciplines are increasingly concerned with correlating physical structure with physical properties. This has been long recognized in the lock-and-key functionality of biological structures such as enzymes and proteins. In condensed matter physics, the properties of many emergent materials, in particular (though not exclusively) transition metal oxides, depend exceedingly sensitively

P. R. Willmott (✉)
Swiss Light Source, Paul Scherrer Institute, 5232 Villigen, Switzerland
e-mail: philip.willmott@psi.ch

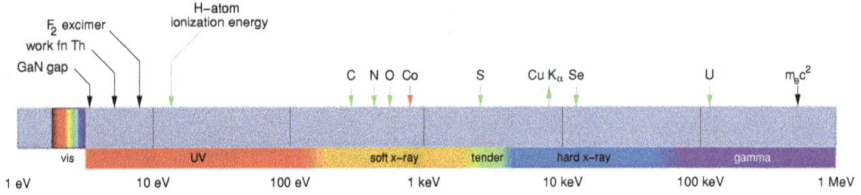

Fig. 1.1 Photon energies and the electromagnetic spectrum. Above the visible regime, spanning only an energy range of 1.77–3.1 eV, the electromagnetic spectrum is divided into UV (up to approximately 140 eV), soft X-rays (up to 2 keV), tender X-rays (to 4 keV), hard X-rays (to 50 keV) and gamma rays (above 50 keV). Important photon energies are highlighted, green (K-edge) and red (L-edge) arrows pointing down imply absorption energies, while those pointing up imply emission lines

on the structure—even changes of a few picometres in bond length, or one degree in bond angle, can have fundamental consequences on the material's electronic character [1].

Protein crystals can often only be grown with linear dimensions of a few tens of micrometres, while interfacial regions between different oxide materials which exhibit unexpected properties [2] may only be a few nanometres thick. Any signal from irradiation of such samples using X-rays, be it the degree of absorption, the amount of elastically scattered radiation, fluorescence or the photoelectron yield, is likely to be very weak. This sets a premium on finding a very intense X-ray source—synchrotrons and X-ray free-electron lasers (XFELs) have been developed to fulfil this need.

The broad range of applications of X-rays has manifested itself in the last two decades in the heterogeneity of scientific fields served and the broad spectrum of techniques now available at synchrotron facilities, representing perhaps the clearest illustration of multidisciplinary research, covering the natural and medical sciences and many aspects of engineering and technology. Nowadays, there are worldwide more than seventy facilities in operation, or under construction, providing sophisticated investigative tools for well over 110 000 users from virtually every field of the natural and engineering sciences.

These users need to understand the operating principles of synchrotrons and their generation of X-radiation in order to best prepare, firstly, proposals to be submitted to the highly competitive review procedure at synchrotrons, and secondly, the beamtimes themselves. This brief overview of synchrotrons, synchrotron sources, and XFELs draws substantially from chapters on synchrotron and XFEL physics in [3] and as such is intended as an accessible primer to the interested reader from any branch of the natural and engineering sciences.

In the next section, the architecture and operating principles of synchrotrons are described and the standard figure of merit for synchrotron light, called the brilliance, is introduced. The three different types of source (bending magnets, wigglers and undulators) are presented in Sect. 1.3. We finish this section by describing ways to control the polarization of X-rays at synchrotrons.

Sections 1.4 and 1.5 outline the most pertinent features of the latest generation of storage rings, so-called diffraction-limited storage rings (DLSRs), and X-ray free-electron lasers (XFELs), respectively.

1.2　A Brief Description of Synchrotrons

1.2.1　Introduction

A synchrotron consists of a ring-shaped evacuated vessel (the storage ring, having a circumference measured typically in a few hundreds of metres, Fig. 1.2) in which high-energy electrons circulate at highly relativistic velocities, and so-called 'beam-lines', that extract and use the radiation emitted by the electrons tangentially to their orbital path, at positions defined by components known as bending magnets (BMs) and insertion devices (IDs).

The electron energy $\mathcal{E}$ at synchrotrons is typically of the order of a few GeV. The emitted photons, on the other hand, have energies measured anywhere between a few eV (just above the visible) and several hundred keV, in the ultrahard X-ray regime.[1] Note, however, that even the photon energies of the latter are still some four orders of magnitude smaller than the electrons' kinetic energy $\mathcal{E}$ in the storage ring.

The electrons are forced into a closed path by bending magnets, which exert a centripetal Lorentz force on them. It is here, and in straight sections in which the insertion devices are installed (see Sect. 1.3), that they emit electromagnetic (EM) radiation.

The electrons lose energy due to their emitting EM radiation. This must be replenished and is achieved via axial acceleration through one or more radio-frequency (RF) cavities installed in the storage ring.

1.2.2　The Lorentz Factor

Before we proceed further, the dimensionless parameter γ is introduced. This so-called 'Lorentz factor' expresses the ratio of the electron energy $\mathcal{E}$ to the rest mass energy of the electrons $m_e c^2 = 511$ keV ($m_e = 9.109 \times 10^{-31}$ kg is the electron rest mass and $c = 2.9979 \times 10^8$ m s^{-1} the speed of light), that is

$$\gamma = \frac{\mathcal{E}}{m_e c^2}. \tag{1.1}$$

---

[1] Some facilities host beamlines that extend down into the far-infrared regime; these relatively uncommon sources are not discussed here.

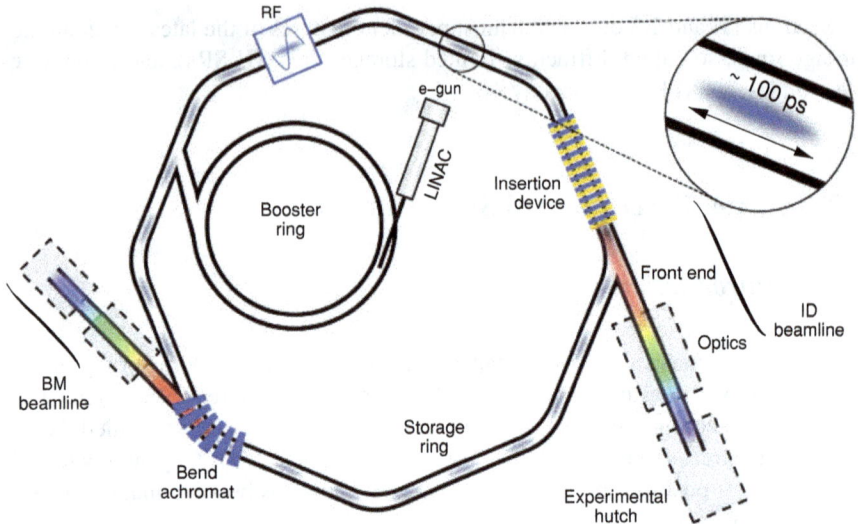

Fig. 1.2 A schematic of the basic components of a modern synchrotron facility. Electrons from a source such as a heated filament in an electron gun are accelerated by a linear accelerator (LINAC) and then injected into a booster ring, where they are further accelerated. They are then further injected into the so-called storage ring. There, they are maintained in a closed path using bending-magnet achromats at arc sections. The beamlines use the radiation emitted from insertion devices (IDs, either wigglers or, more commonly, undulators) placed at the straight sections between the arcs, and from the bending magnets (BM), on the axes of emission. The energy lost by the electrons through the emission of synchrotron light is replenished by particular parts of the cycle of one or more radio frequency (RF) supplies. This forces the electrons within the storage ring to separate into discrete bunches. Each bunch contains of the order of 10^9 electrons and has a full width at half maximum duration of the order of 100 ps.

Consequently, for typical storage ring energies of the order of a few GeV, γ is of the order of a few thousand to a little over 15 000 (for the Advanced Light Source in Berkeley, $\mathcal{E} = 1.9$ GeV and hence $\gamma = 3718$, while the highest energy storage ring, SPring8, has a storage ring energy of $\mathcal{E} = 8$ GeV, leading to $\gamma = 15\,656$). The Lorentz factor crops up in many equations related to synchrotron radiation (SR), including the beam divergence, relativistic electron mass, electron emittance and the radiative power output.

From the special theory of relativity, it emerges that

$$\gamma = \left[1 - (v/c)^2\right]^{-1/2}, \tag{1.2}$$

where v is the electron's velocity. We re-arrange this to obtain

$$v = c\left(1 - 1/\gamma^2\right)^{1/2}. \tag{1.3}$$

We know γ is of the order of several thousand, hence $1/\gamma^2$ is a very small number, of the order of 10^{-8}. We use the approximation $(1-x)^n = 1 - nx$ for $x \ll 1$ to obtain

$$\frac{c-v}{c} \approx \frac{1}{2\gamma^2}. \tag{1.4}$$

In other words, the difference between c and v is very small, of the order of a few $m\,s^{-1}$.

Lastly, the mass of the electron from the perspective of a stationary observer is equal to γm_e. We will return to these findings later.

1.2.3 Dipole Radiation and Synchrotron Radiation

Why do electrons emit EM radiation at all? First, it should be stressed that electrons, or indeed any charged particles, only emit EM radiation when accelerated. 'Accelerated' can include the conventional meaning of an increase in speed but no change in direction; its opposite, that is, a deceleration; or a change in the electrons' direction, such as in centripetal acceleration. The second case corresponds to the so-called 'Bremsstrahlung' (German expression for braking radiation), while the third is associated with SR.

Given this, why then does the action of accelerating electrons cause them to emit light? Consider Fig. 1.3. EM radiation is a form of *transverse* wave in which the oscillations (of the electric and magnetic fields) are at right angles to the direction of motion. For the sake of simplicity, we consider here only the electric field component of the EM radiation. The electric field lines of the electrostatic field of a stationary and isolated electron emanate out radially from the electron. Any observer looking at the electron, therefore, sees no transverse component of the field, which thus implies she sees no radiation [Fig. 1.3a].

If, however, the electron is made to execute oscillatory motion, the electric field lines, which are anchored to the electron and emanate out from it at the speed of light, will reflect this motion and thus also oscillate accordingly [Fig. 1.3b]. In all directions except that exactly along the axis of acceleration, our observer will 'see' a transverse component to the electric field and therefore perceive that light is being emitted. The amplitude of the radiation is proportional to $\cos \chi$, where χ is the polar angle between the axis of acceleration and the observer's direction. The intensity distribution, shown in Fig. 1.3b, is proportional to $\cos^2 \chi$. This so-called 'dipole radiation' is the reason why mirrors reflect visible light, radio antennae capture or emit radio waves and undulators produce X-radiation.

SR is highly collimated, with natural divergences of the order of 0.1 mrad (mrad; 1 mrad is approximately equal to 0.06°). A detailed derivation of the spatial distribution of SR lies beyond the brief of this introductory overview (see, for example, [3]). Suffice it to say, it differs substantially from the dipole distribution shown in Fig. 1.3b; this is due to relativistic Doppler shifting. The angular power distribution

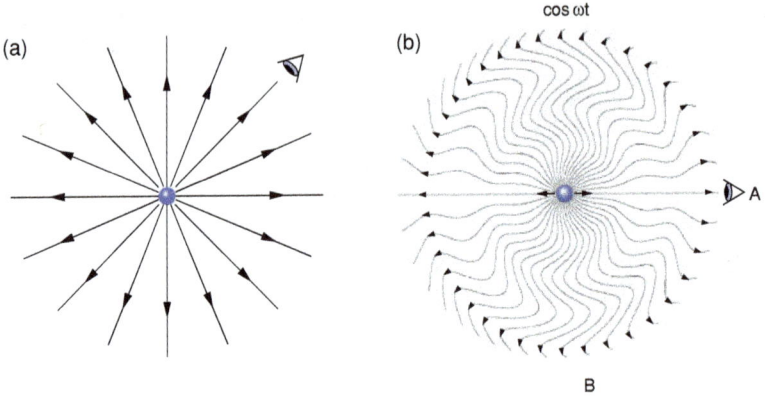

Fig. 1.3 Generation of EM radiation through the acceleration of charged particles. **a** A charged particle at rest or moving at a constant velocity will not emit light, as an observer of the particle will detect no lateral component of the electric field lines. **b** If, however, the particle is accelerated, an observer positioned anywhere except along the axis of that acceleration (position A) will experience a shift in the position and direction of the electric field lines as the event horizon washes over them at the speed of light (for example, at position B). A simple harmonic driving force will generate radiation from the electrons at the same frequency.

per solid angle $dP/d\Omega$ for an electron moving with a velocity v is given without derivation as

$$\frac{dP}{d\Omega} = \kappa \frac{a^2}{(1 - \beta \cos\theta)^3} \left[1 - \frac{\sin^2\theta \cos^2\phi}{\gamma^2(1 - \beta \cos\theta)^2} \right] \tag{1.5}$$

where ϕ and θ are the polar (out of the orbital plane) and azimuthal (in the orbital plane) angles, respectively, $\kappa = e^2/(16\pi^2\epsilon_0 c) = 6.124 \times 10^{-38}$ kg m^2 s^{-1} and $a = Bec/(\gamma m_e)$ is the acceleration perpendicular to the direction of motion due to the Lorentz force exerted on the relativistic electron by the magnetic field B. Note that in the nonrelativistic limit of $v \ll c$ ($\beta = v/c \ll 1$), (1.5) reduces to the simple $\cos^2\theta$ dependence of dipole radiation. The progression from dipole to synchrotron radiation for different values of β is shown in Fig. 1.4).

For a given centripetal acceleration a, the maximum power (in the forward direction) scales with the fourth power of the electron-beam energy. At highly relativistic velocities, it emerges that

$$\frac{dP}{d\Omega} = \kappa' B^2 \gamma^4 \frac{\left[1 - (\gamma\theta)^2\right]^2}{\left[1 + (\gamma\theta)^2\right]^5} \tag{1.6}$$

where $\kappa' = e^4/(2\pi^2 m_e^2 \epsilon_0 c) = 1.5156 \times 10^{-14}$ m^2 C^2 kg^{-1} s^{-1}) (or W T^{-2}). In the frame of reference of the electron, the distribution remains pure dipole radiation

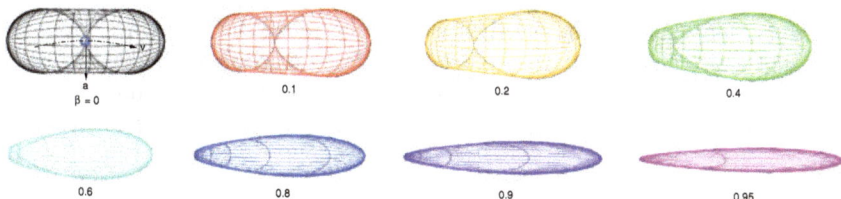

Fig. 1.4 From dipole radiation to synchrotron beams. The progression of the angular power distribution of radiation from an electron travelling at a fraction of the speed of light $\beta = v/c$ while experiencing a centripetal acceleration a perpendicular to its motion. The case $v = 0$ corresponds to dipole radiation. As β increases, the radiation distribution is swept in the forward direction

Fig. 1.5 Plot of the exact (blue solid curve) and approximate (red dashed curve) expressions given in (1.8), as a function of β up to $\beta = 0.9999$. More typical values of β at modern synchrotrons are $1 - 10^{-8}$ (0.999 999 99)

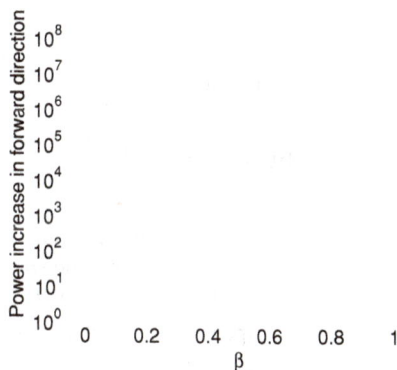

($\beta = 0$ in Fig. 1.4) and the opening angle is $\theta' = \pm\pi/2$. From the perspective of a stationary observer, however, the angular distribution is modified such that

$$\theta = \sin^{-1}\left[\frac{\sin\theta'}{\gamma(1 + \beta\cos\theta')}\right]. \tag{1.7}$$

Thus, the opening angle changes from $\pm\pi/2$ in the electrons' frame of reference to $\pm 1/\gamma$ in the laboratory frame. The entire beam therefore lies within $\pm 1/\gamma$, and has a full width at half maximum of approximately $1/\gamma$. This 'natural opening angle' (or divergence) of the narrow radiation cone, for typical storage ring energies of $1-8$ GeV, is equal to $0.5-0.06$ mrad ($0.028-0.0034°$), respectively; SR is highly collimated.

Lastly, it can be simply demonstrated from (1.2) and (1.5) that, for a given acceleration a, the ratio of the power in the forward direction ($\theta = 0$) for an electron beam travelling at a non-zero speed perpendicular to a, to that of the maximum of dipole radiation for an electron with zero velocity perpendicular to the acceleration is

$$\frac{P(\theta = 0, \beta \neq 0)}{P(\theta = 0, \beta = 0)} = \frac{(1 - \beta^2)}{(1 - \beta)^3} \approx 8\gamma^4 \tag{1.8}$$

where the approximation on the right is valid for relativistic velocities. Both the exact and approximate expressions are shown in Fig. 1.5 up to $\beta = 0.9999$. Increases in power of the order of 10^{16} compared to electrons undergoing stationary dipole oscillations can thus be expected—synchrotrons truly do deliver powerful beams!

1.2.4 Spectral Flux, Emittance, and Brilliance

Flux and brilliance are figures of merit of the quality of a synchrotron facility. The spectral flux is defined as the number of photons per second per unit bandwidth (BW), normally given as 0.1%, and is the appropriate measure for experiments that use the entire, unfocussed X-ray beam. Brilliance essentially states how tightly the spectral flux is collimated and how small the source size is. It is defined as

$$\text{Brilliance} = \frac{\text{photons/second}}{(\text{mrad})^2 \, (\text{mm}^2 \text{ source area}) \, (0.1\% \text{ BW})} \, , \tag{1.9}$$

and is therefore equal to the flux per unit source cross-sectional area and unit solid angle. Note that the flux (as against the *spectral* flux) is simply measured in photons per second. Doubling the transmitted BW from a broadband source thus doubles the flux, but leaves the spectral flux unchanged. From (1.9), it is seen that the brilliance is inversely proportional to both the source size and the beam divergence. The product of the linear source size σ and the beam divergence σ' in the same plane is known as ϵ, the emittance in that plane, that is

$$\epsilon_x = \sigma_x \, \sigma_x' \, , \tag{1.10}$$

$$\epsilon_y = \sigma_y \, \sigma_y' \, . \tag{1.11}$$

The goal of the machine physicist designing a synchrotron magnet lattice is to provide as low an emittance as possible, in other words, a source with an exceedingly small cross section emitting X-rays that are highly collimated. For a given synchrotron storage ring, the emittance in each transverse direction (x, in the orbital plane, or y, perpendicular to the orbital plane) is, according to Liouville's theorem, a constant. The emittance will be different for different facilities, in each case being determined primarily by the degree of sophistication and perfection of the magnet lattice.

Importantly, the total emittance in a given plane is a convolution of the contribution from the electron beam and that from the emitted photons. It follows that the total source size $\sigma_{x,y}$ and divergence $\sigma_{x,y}'$ in the x- and y-planes perpendicular to the direction of propagation of a given storage ring are also convolutions of contributions from the electron and photon beams, that is

$$\sigma_{x,y} = \left[(\sigma_{x,y}^e)^2 + (\sigma^P)^2 \right]^{1/2} \, , \tag{1.12}$$

$$\sigma_{x,y}' = \left[(\sigma_{x,y}'^e)^2 + (\sigma'^P)^2 \right]^{1/2} \, . \tag{1.13}$$

While the electron contribution can be minimized by sophisticated electron optics (see Sect. 1.4), the photon emittance is an intrinsic property defined by Heisenberg's uncertainty principle, and is equal to $\lambda/4\pi$ in both the x- and y-plane. In third-generation storage rings, the electron emittance in the orbital plane ϵ_x^e dominates the total emittance and is thus the limiting factor for the brilliance. In fourth-generation DLSRs, the benchmark for modern storage ring designs, ϵ_x^e has been reduced to values close to or even below (in the case of soft X-rays) the intrinsic photon emittance, which we consider in detail in Sect. 1.4. In other words, the emittance is no longer limited by the electron optics, but by the fundamental optical diffraction limit. This is the meaning of the moniker 'diffraction-limited storage ring' defining the fourth-generation synchrotron facilities now coming online.

Synchrotrons have brilliances using modern undulators of approximately 10^{22} photons s^{-1} mrad^2 mm^{-2} 0.1% bandwidth^{-1}. This is some 12 orders of magnitude higher than that of a standard laboratory-based Cu $K\alpha$ source and less than a factor of 100 lower than high-quality visible laser sources. The main reasons for this are the size of the radiation source, of the order of ten micrometres at fourth-generation DLSRs, the high collimation of the beams, being of the order of 10 μrad in the orbital plane and the fact that synchrotrons emit an enormous amount of light. The power emitted by an electron is proportional to the square of the electron's centripetal acceleration a, and the fourth power of the storage ring energy.

1.2.5 The Radio-Frequency Power Supply

Conservation of energy dictates that the kinetic energy of the electrons is dissipated due to emission of radiation at the bending magnets and insertion devices. This energy must be replenished, or otherwise, the electrons would spiral into the inner wall of the storage ring. This is achieved by boosting the electrons' energy at one or more positions along the storage ring as they pass through RF cavities [Fig. 1.6a]. This requires that the electrons enter the cavity at a certain point of the RF cycle.

Because the electrons can only receive the correct amount of energy at very specific and narrowly defined values in the RF cycle, they are separated into a series of packets, or 'bunches'. The energy loss of the electrons for each cycle around the ring is given by the total power loss of the storage ring divided by the storage ring current and is equal to approximately 0.2–1 MeV, or of the order of 0.05% of the nominal electron energy. Depending on the size of the facility, most storage rings host between two and eight RF cavities. Between them, they must be able to replenish this loss.

Consider Fig. 1.6b. On average, the electrons require a certain energy boost in order to keep them on a stable path, given by an amount eV_{ref}. If an electron loses more than this amount of energy, it will enter the RF cavity somewhat earlier at point A. This might sound counterintuitive—surely if the electron has less energy, it will be slower and enter the cavity later. But because it takes a shorter path in the bends according to the linear dependence of the bending radius and the electron energy it

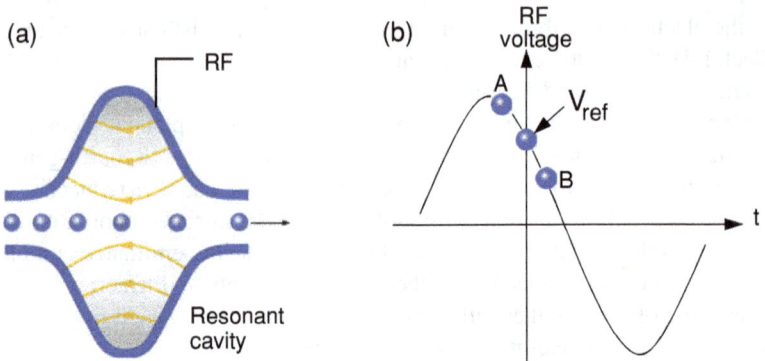

Fig. 1.6 Replenishing the electron energy in a storage ring. **a** Electrons entering the resonant RF cavity at the correct moment in its voltage cycle are accelerated by a suitable amount by the electric field within the cavity. Note that the field lines point in the opposite direction to the acceleration, as the electric force is $\mathbf{F_E} = q\mathbf{E}$ but an electron has a negative charge $-e$. **b** 'Slow' electrons entering the RF cavity at A will be given more of a boost than 'fast' electrons at B.

indeed will arrive earlier than a higher energy electron, and will thus experience a larger acceleration than if it were at the reference voltage. Likewise, if the electron is too fast, it will receive less of a boost. Any electrons entering the RF cavity outside this narrow range above and below the reference voltage will not gain the correct energy and will be lost to the system. The electrons therefore quickly bunch into packets associated with each cycle of the RF cavity.

The short bunch lengths allow users to exploit the time structure of SR down to well below the nanosecond time scale for time-resolved experiments. These types of experiment became increasingly important in third-generation synchrotron facilities, in areas as diverse as molecular biology, catalysis, condensed matter physics and domain flipping in nanomagnetism and, for DLSRs, they promise to be complementary to XFEL investigations.

1.2.6 Radiation Equilibrium

What determines the electron emittance? The emittance of a storage ring is determined by the opposing influences of two phenomena: radiation damping (something you want) and quantum excitation (something you don't). At the NSLS II in Brookhaven, the machine performance is optimized by maximizing radiation damping, while in the next generation of DLSRs such as MAX-IV, quantum excitation has been minimized. As we have already stated, radiation damping improves the emittance by reducing the transverse momentum component [see Fig. 1.7a]. When an electron emits a photon, it loses the energy of the photon. This causes it to oscil-

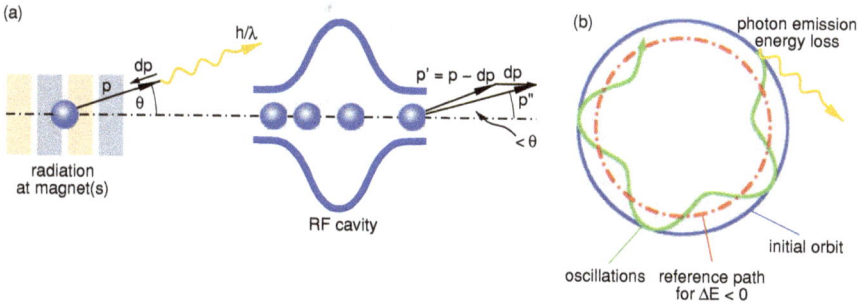

Fig. 1.7 Radiation equilibrium between quantum excitation and radiation damping. **a** Radiation damping. An electron traversing a magnet in an insertion device is made to deviate from the central axis by an angle θ, due to the Lorentz force. Emission of a photon with momentum $h/\lambda = \hbar k$ will be in the direction of the electron at that instant in time. Conservation of momentum dictates that the electron's momentum will be reduced to $p' = p - h/\lambda$. The same electron will regain this momentum loss dp after travelling through the RF cavity; importantly, this will now be parallel to the central axis, thus reducing the angle of the electron's momentum p'' to the central axis and hence also the electron beam's emittance. **b** Quantum excitation. An electron loses energy due to the emission of a photon and begins to oscillate around a new reference orbital path with a smaller radius. These oscillations induce a stochastic distribution of transverse momenta, thereby increasing the emittance. Reproduced from [3] with permission

late around a new reference orbit, thus broadening the beam and thereby increasing the emittance [Fig. 1.7b]. Moreover, the dispersion of the electron beam increases. Quantum excitation can be reduced by designing the magnet lattice so that the electron energy dispersion is minimized at the main locations of radiation, namely the bending magnets. This is achieved by horizontal focusing at the bends and the use of many small deflection angle bends in multibend achromat lattices (see Sect. 1.4) to limit dispersion growth.

1.2.7 Coherence

We now consider coherence, including both longitudinal and transverse coherence. The latter depends on the source size and divergence of the photon beam, in other words, it depends on the emittance; longitudinal coherence depends on the bandwidth. The coherent fraction of a beam is critically important in lensless imaging techniques and photon correlation spectroscopy, plus also in phase-contrast tomography. Note also that bound up in the figure of merit of brilliance are the above parameters that quantitatively define coherence: the emittance and the relative spectral BW. Figure 1.8 provides a schematic summary of coherence. Brilliance really does encompass the most important qualities of synchrotron light; because however, it combines flux, spatial coherence and longitudinal coherence, it is important to

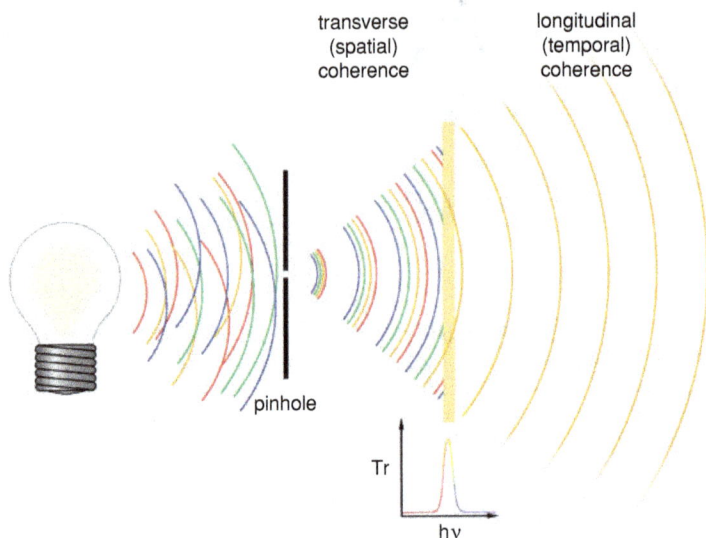

Fig. 1.8 Coherent radiation. The coherent fraction of a broadband, spatially extended, source can be selected as follows. Firstly, a pinhole selects a small, spatially constrained fraction of the radiation thereby acting as a secondary, quasi-pointlike, source. This secondary source has an improved transverse (or spatial) coherence, as the emittance, which is the product of the radiation's divergence and source size, is now much smaller. Next, a filter (which is normally a monochromator) selects a narrow BW which is much narrower than the original source. Now, the radiation is spatially *and* longitudinally (or temporally) coherent. Note that both the emittance and relative spectral BW are included in the definition of brilliance.

know which factors determine a given value. For example, is the brilliance high because there are more photons on the sample, or because the emittance is small?

No X-ray source has an infinitely narrow bandwidth. Consequently, the different frequency components within the beam will sooner or later drift out of phase with one another. The time for the phase between two waves differing in frequency by an amount $\Delta \nu$ but which are initially in phase to differ by π radians (i.e. from fully constructive to fully destructive) is simply $1/2\Delta \nu$. This is known as the so-called longitudinal coherence time, $\Delta \tau_c^{(l)}$. During this time, the waves have travelled in vacuum a distance $l_c^{(l)} = c\Delta \tau_c^{(l)}/2$, known as the longitudinal (or temporal) coherence length [see Fig. 1.9a], given by

$$l_c^{(l)} = \frac{\lambda^2}{2\Delta\lambda}.$$

(1.14)

The longitudinal coherence after a monochromator is usually determined by the rocking curve of the crystal or grating used in the monochromator, which defines $\lambda/\Delta\lambda$. For a perfect crystal with insignificant mosaicity, $\Delta\lambda$ is limited by the so-called Darwin width, and $\lambda/\Delta\lambda$ can easily exceed 10^4—the relative bandwidth of a Si(111) single crystal is determined by the extinction depth and is approximately 1.4×10^{-4}. The longitudinal coherence length can thus be several micrometres, even in the hard X-ray regime.

The *transverse coherence length* $l_c^{(t)}$ (also called the spatial coherence length) results from the interference of waves having the exact same wavelength but with slightly different directions of propagation. This arises because all sources have a finite size D and a non-zero divergence $\Delta\theta$ (that is, a non-zero emittance), as shown in Fig. 1.9b. In this case,

$$l_c^{(t)} = \lambda/2\Delta\theta = \lambda R/2D , \qquad (1.15)$$

where D is the linear size of the finite source and R is the distance from the source to the observation point. If we assume the source has a Gaussian profile, determination of D requires integration of interference contributions across the entire source's intensity distribution. It emerges that $l_c^{(t)}$ is related to the standard deviation of the beam size $\sigma_{x,y}$ by

$$l_c^{(t)} = \lambda R/(2\pi^{1/2}\sigma_{x,y}) , \qquad (1.16)$$

or, in practical units

$$l_c^{(t)}[\mu m] = 28.21 \frac{\lambda \, \mathring{A} \, R \, [m]}{\sigma_{x,y} \, [\mu m]} . \qquad (1.17)$$

Note that the transverse coherence length can be made larger by judicious use of slits limiting the apparent source size and divergence, but obviously at the cost of flux. Beamlines such as coherent lensless imaging beamlines tend to be very long in order to maximize R.

In the vertical direction, the source size at an undulator of, say, 2-m length, is of the order of $\sigma_y = 2 \, \mu m$. For 1-Å radiation, this yields a vertical spatial coherence for an observer at 40 m of $l_c^{(t,y)} = 564 \, \mu m$. The horizontal spatial coherence has traditionally been two orders of magnitude smaller than this, due to the very much larger extent of the electron beam in the orbital plane. The electron beam source size in DLSR storage rings in the horizontal direction may, however, be as small as $\sigma_x = 10 \, \mu m$; the corresponding coherence length $l_c^{(t,x)}$ is of the order of 0.1 mm.

1.3 Sources of Synchrotron Radiation

In this section, the three sources of SR are semi-quantitatively described, while the differences between the radiation produced by bending magnets and wigglers on the one hand, and undulators on the other, are presented in a heuristic manner.

(a)

(b)

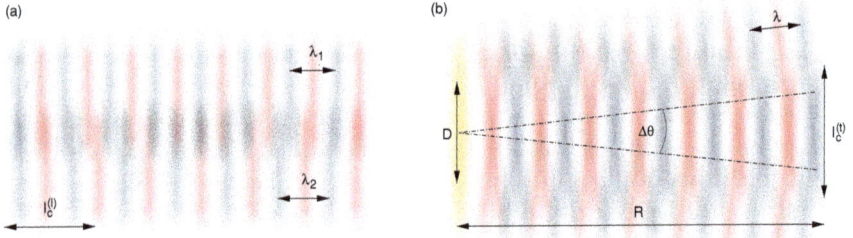

Fig. 1.9 Beam coherence. **a** The temporal, or longitudinal, coherence length is determined by the monochromaticity of the source, while **b** the transverse, or spatial, coherence length depends on the photon beam emittance.

1.3.1 Bending Magnets and Wigglers

As mentioned above, the electrons are maintained on a closed path via magnetic, or Lorentz, forces. The Lorentz force, $\mathbf{F}_L$, is proportional to the cross-product of the magnetic field strength, $\mathbf{B}$, and the charged particle's velocity, $\mathbf{v}$, that is

$$\mathbf{F}_L = e\mathbf{B} \times \mathbf{v} , \tag{1.18}$$

where $e = 1.6022 \times 10^{-19}$ C is the elementary charge. It acts perpendicular to the plane defined by $\mathbf{B}$ and $\mathbf{v}$. For the sake of simplicity, we only consider those cases where $\mathbf{F}_L$, $\mathbf{B}$, and $\mathbf{v}$ are mutually orthogonal, and drop the bold face implying their vectorial nature. Now,

$$F_L = eBv . \tag{1.19}$$

We equate this with a centripetal force mv^2/ρ, where $m = \gamma m_e$ is the relativistic mass of the electron travelling at a speed $v \approx c$. The bending radius of the centripetal force, ρ, is equal to the bending magnet radius. Therefore, to a high degree of accuracy,

$$eBc \approx \frac{\gamma m_e c^2}{\rho} \tag{1.20}$$

$$\Rightarrow \rho = \frac{\mathcal{E}}{ceB} . \tag{1.21}$$

In practical units, we obtain

$$\rho = 3.3 \frac{\mathcal{E}[\text{GeV}]}{B[\text{T}]} . \tag{1.22}$$

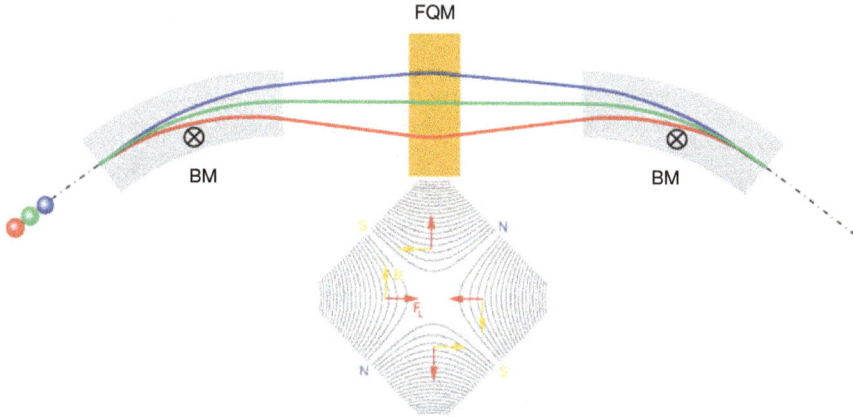

Fig. 1.10 Double-bend achromats. The angular dispersion of electrons of different energies as they pass through a bending magnet and the consequent increase in the electron beam emittance can be corrected in a DBA by placing a focusing quadrupole magnet (FQM) symmetrically in between two identical bends (BM). Lower energy electrons are bent through larger angles than are those of higher energy.

For typical storage ring energies of a few GeV and magnetic field strengths of the order of 1 T, we obtain bending magnet radii of the order of 10 m.

The relative spread in electron energy $\Delta \mathcal{E}/\mathcal{E}$ in a storage ring is of the order of 10^{-3}. The bending radius ρ is directly proportional to the electron energy [see (1.21)]. Therefore, the path of those electrons with more (less) than the central energy will have a larger (smaller) radius, resulting in an unwanted increase in emittance. This problem is resolved by using an arrangement of bending and quadrupole magnets is known as a double-bend achromat (DBA, also called a Chasman–Green lattice, after its inventors [4]), shown in Fig. 1.10.

The natural (i.e. minimum) horizontal electron emittance of a DBA with bending angle 2θ (that is, θ for each dipole pair) is given by

$$\epsilon_{x,\text{DBA}} = C_{\text{DBA}} \gamma^2 \theta^3 \,, \tag{1.23}$$

where $C_{\text{DBA}} = 11\sqrt{5}\hbar/384\,m_e c = 2.474 \times 10^{-5}$ nm [5]. So, for example, the lower limit emittance of a 3 GeV storage ring containing 20 DBAs would be 3.3 nm rad, larger by nearly three orders of magnitude than the photons' diffraction-limited value of $\lambda/4\pi = 8$ pm rad calculated for 1-Å radiation. Efforts to approach the diffraction limit by using multibend achromats (MBAs) are discussed in Sect. 1.4.

The primary purpose of bending magnets is to maintain the electrons in the storage ring on a closed path. Bending magnets have typical magnetic field strengths of the order of 1 T. They produce bending magnet radiation in a flattened cone with a fan angle equal to that swept out by the path of the electrons due to the Lorentz forces they are subjected to. The relatively large subtended angle of bending magnet

Fig. 1.11 $K_{2/3}(x)$, the modified Bessel function of the second kind for order $2/3$

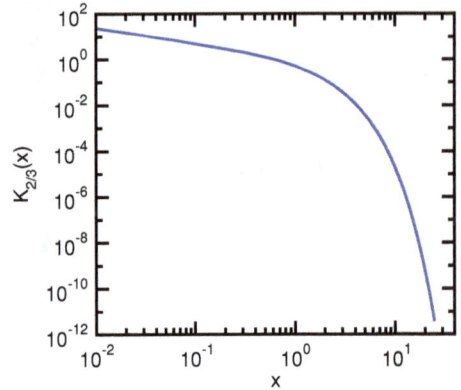

Fig. 1.12 Bending -magnet spectra for on-axis horizontally polarized radiation for three different combinations of the storage ring energy and magnetic field strength.

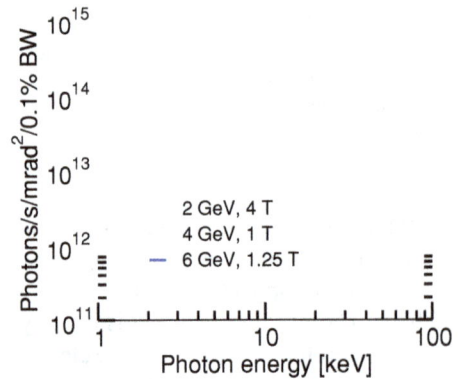

radiation, measured in degrees, allows one to accommodate more than one so-called 'bending-magnet beamline' at a single bending magnet.

The spectral flux distribution is given by

$$\frac{\text{ph/s}}{(\text{mrad})^2\, 0.1\%\,\text{BW}} = 1.325 \times 10^{10} I_e[\text{mA}]\mathcal{E}^2[\text{GeV}]\left(\frac{E}{E_c}\right)^2 K_{2/3}\left(\frac{E}{2E_c}\right), \quad (1.24)$$

where $E_c = 3\hbar e B\gamma^2/2m_e$ is the so-called critical, or characteristic, energy (which in practical units can be expressed as $E_c[\text{eV}] = 665.023\, B[\text{T}]\, \mathcal{E}^2[\text{GeV}^2])$, and $K_{2/3}(x = E/2E_c)$ is the modified Bessel function of the second kind for nonintegral order (in this instance, $2/3$), shown in Fig. 1.11.

The spectral flux is thus determined by the storage ring energy and the magnetic field strength. Increasing B (but keeping $\mathcal{E}$ constant) shifts the maximum of the spectrum to higher photon energies but does *not* increase the spectral flux's maximum. In contrast, only increasing $\mathcal{E}$ both shifts the spectral maximum to higher photon energies and higher values.

The broadband bending magnet spectra for three different combinations of storage ring energy and magnetic field strength are shown in Fig. 1.12. Particularly at low- and medium-energy facilities, the maximum values of both these quantities are too small to extend the spectrum far into the hard X-ray regime. However, if superconducting magnets with larger magnetic field strengths are employed, the photon energy range can be extended to harder X-radiation, as the critical energy is proportional to the magnetic field strength. Moreover, the radiative power increases with the square of the magnetic field strength B. These so-called 'superbends' can have magnetic field strengths as high as 8 T.

There are two types of insertion devices, distinguished from each other by the amount that the electrons are forced to deviate in a slalom-like path from a purely straight path. This at first seemingly subtle distinction has a fundamental effect on the nature of the radiation, however. For angular excursions substantially larger than the synchrotron radiation's natural opening angle γ^{-1}, the radiation cones from each magnet in the insertion device do not overlap. Under these conditions, the intensities produced from each dipole are added and the ID is referred to as a *wiggler*, which is briefly described below.

For gentler excursions of the order of γ^{-1}, the ID is called an *undulator*, described in Sect. 1.3.2.

The maximum angular deviation ϕ_{max} of the electron oscillations in an ID is defined by the dimensionless 'magnetic deflection parameter' K, given by

$$\phi_{max} = K/\gamma . \tag{1.25}$$

K can be expressed in terms of the maximum magnetic field B_0 as

$$K = \frac{e B_0}{m_e c k_{u,w}} = 0.934 \, \lambda_{u,w}[\text{cm}] \, B_0[\text{T}] , \tag{1.26}$$

where λ_u or λ_w are the periods of the oscillations in the undulator or wiggler, respectively, and $k_{u,w} = 2\pi/\lambda_{u,w}$. For a wiggler, K is typically between 10 and 50, while for undulators, K is close to unity and changes according to the size of the gap between the upper and lower magnet arrays. The horizontal spread in the electron beam divergence is

$$\theta_x = 2K/\gamma . \tag{1.27}$$

So, for example, a wiggler having $K = 20$, operating in a 4 GeV storage ring would have a horizontal divergence of 5.2 mrad (0.30°).

A wiggler can be thought of as being a series of bending magnets within a straight section of the storage ring that turns the electrons alternately to the left and to the right. The maximum angular excursion from the central axis is larger than the natural opening angle of the radiation, γ^{-1}. For each oscillation, the electrons are twice moving parallel to (and in reality also very close to) this axis. The radiation is therefore enhanced by a factor of $2N$, where N is the total number of wiggler periods and is of

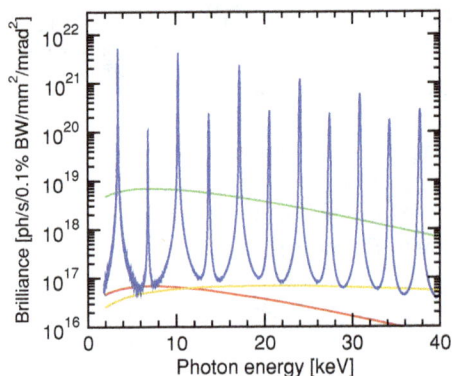

Fig. 1.13 Comparison of brilliances at a 3 GeV DLSR running at 400 mA between a U14 undulator with $K = 1.6$ (blue), a bending magnet with $B = 1.41$ T (red), a superbend with $B = 4$ T (yellow), and a wiggler with the same field strength as the bending magnet and 100 periods (green). Note that the peak brilliance of synchrotron sources can be calculated from the average brilliance shown in this figure by multiplying by the ratio of the pulse separation $\Delta t \sim 5$ ns to the pulse width $\Delta \tau \sim 50$ ps, that is, approximately a factor 100. The peak brilliance of XFELs is of the order of 10^{34} ph s^{-1} mm^{-2} mrad^{-2} 0.1% BW^{-1}, some ten orders of magnitude greater than that produced by DLSR undulators.

the order of 20. The spectrum from a wiggler has the same form as that from a bending magnet—it is broadband and thus produces a large amount of integrated radiative power, of the order of several kW. Thermal management of optical components is thus critical. Wigglers are therefore becoming fairly uncommon, particularly in fourth-generation DLSRs.

1.3.2   Undulators

In undulators, the angular deviation of the electrons away from the central axis is of the order of $1/\gamma$; the radiation cones emitted by the electrons thus overlap as they execute the slalom motion. Consequently, radiation from the dipoles interferes with one another. As such, the field *amplitudes* are added vectorially (i.e. including the phase difference from each contribution) and the sum of this is squared to produce the intensity, which peaks at those wavelengths where interference is constructive.

Undulators therefore differ fundamentally from bending magnets and wigglers in that their spectral flux reflects this interference phenomenon and is hence concentrated in evenly separated, narrow bands of radiation (Fig. 1.13). The first practical undulator device to operate in the X-ray regime was constructed by Klaus Halbach and co-workers at the Lawrence–Berkeley National Laboratory and tested at the SSRL synchrotron at Stanford in 1981. This breakthrough was thanks on the one hand to the development of novel magnetic alloys such as SmCo$_5$ [6], allowing the construction of magnet arrays with the required small periodicities and high magnetic field strengths [7]; and on the other, to a clever arrangement of pole orientations (referred to as the 'Halbach array') which effectively suppresses the field strength on one side of the array and doubles it on the other, thus maximizing the magnetic flux where it is needed.

The four basic parameters for undulator radiation from a device of length L are the relativistic Lorentz factor γ, the undulator spatial period λ_u, the number of periods in the magnet array $N = L/\lambda_u$, and K. As already stated, for an undulator, K is about unity. K can be varied by changing the gap size between the upper and lower arrays of magnets; this tunes the spectrum so that a suitable near-lying spectral maximum sits at the desired photon energy. The transformation from wiggler to undulator radiation is achieved in practice not by reducing the lateral excursions through reduction of the magnetic field strength between the magnetic pole pairs—this would result in an unacceptable drop in flux—but instead by reducing the magnetic pole spatial periodicity λ_u [see (1.26)].

It emerges that the condition for constructive interference is given for the mth harmonic by

$$m\lambda_m = \frac{\lambda_u}{2\gamma^2}\left(1 + \frac{K^2}{2}\right),$$ (1.28)

or in practical units

$$m\lambda_m \overset{\circ}{\mathrm{A}} = \frac{13.056\,\lambda_u[\mathrm{cm}]}{\mathcal{E}^2[\mathrm{GeV}]}\left(1 + \frac{K^2}{2}\right).$$ (1.29)

The intrinsic source size and divergence of undulators associated exclusively with photon emission (i.e. ignoring the electron emittance) are given by

$$\sigma^P = \frac{1}{4\pi}\sqrt{\lambda L}$$ (1.30)

and

$$\sigma'^P = \sqrt{\frac{\lambda}{L}},$$ (1.31)

resulting in an intrinsic emittance of

$$\epsilon^P\,[\mathrm{pm\,rad}] = \frac{\lambda}{4\pi} = \frac{98.66}{E[\mathrm{keV}]}.$$ (1.32)

Note that the divergence σ'^P can also be expressed in terms of the harmonic number m and number of periods N, and is approximately equal to $\sigma'^P = 1/(\gamma\sqrt{mN})$. The

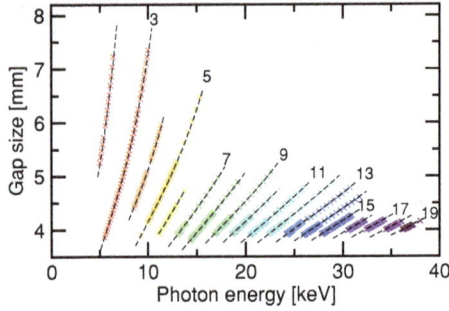

Fig. 1.14 Measurements of optimized gap positions for different photon energies and utilized harmonics of the U14 undulator of the Materials Science beamline, SLS, from the second to nineteenth harmonic. In normal operation, scanning to higher energies within any given harmonic is achieved by opening the undulator gap (the progression of this between the third and fifth harmonics is shown as the dot-dashed blue lines). One moves from a given harmonic to the next higher (red dot-dashed lines) when the desired photon energy can be accessed at the higher harmonic with a gap size no smaller than the minimum allowed value, here 4 mm, shown as the blue horizontal dashed line.

divergence thus becomes smaller with harmonic number and number of periods in the undulator.

The interference spectrum at an angle θ away from the central axis of the undulator is shifted towards lower energies, and is given by

$$m\lambda_m(\theta) = \frac{\lambda_u}{2\gamma^2}\left(1 + \frac{K^2}{2} + \gamma^2\theta^2\right). \qquad (1.33)$$

For example, an observer positioned off axis by half the natural opening angle $\theta = 1/2\gamma$ (approximately 3.5 mm at a distance of 40 m for a facility with a 3 GeV storage ring energy) would, for $K = 1$, see a spectrum shifted by a factor 7/6 to the red.

The spectral width of the undulator harmonics is inversely proportional to the number of periods, N. As in any interference or diffraction set-up, the condition for constructive interference becomes increasingly strict the larger the number of involved 'scatterers' (in this instance, the $2N$ magnet pairs). Hence, the inverse of the relative bandwidth, called variously the monochromaticity or the quality factor $\lambda_m/\Delta\lambda_m = \nu_m/\Delta\nu_m$, is equal to the number of periods N multiplied by the harmonic number m. As an example, the tenth harmonic of an undulator consisting of $N = 70$ periods has a relative bandwidth $\Delta\lambda_m/\lambda_m$ of 1.4×10^{-3}.

The undulator spectrum is tuned by varying K. This is achieved by changing the gap between the two sets of magnetic poles and thereby the magnetic field strength B_0 [see (1.26) and Fig. 1.14].

Note that, for reasons of symmetry, even harmonics are in general weaker than odd harmonics. Higher K undulators provide both more intense higher-energy harmonics than lower K devices, while the difference in intensities between even and odd harmonics is less pronounced.

1.3.3   Polarization of Synchrotron Radiation

Polarization of light in spectroscopy is a highly valuable tool, as certain selection rules imply that certain electronic features can be distinguished merely by using differently polarized X-radiation, for example, in the study of surface absorbates on crystalline surfaces, or in distinguishing between differently oriented ferromagnetic or antiferromagnetic domains. Magnetic dichroism using tunable and polarized soft or hard X-rays at synchrotron sources offers the unique feature of chemical specificity. Magnetic features down to below the 100-nm scale can be imaged by exploiting the dependence of the absorption on the polarization of X-rays in ferromagnetic and anti-ferromagnetic materials. The methods are called X-ray magnetic circular dichroism (XMCD) and X-ray magnetic linear dichroism (XMLD), respectively, for ferromagnetic and antiferromagnetic materials (see Chap. 4). Imaging down to the nanometre range is very important for magnetic structures, as at this scale, the influence of domain boundaries between one magnetic direction and another becomes significant and new phenomena can occur which would be negligible in larger structures. An understanding of the energetics of nanomagnetism is thus essential in the drive to further miniaturize magnetic memory storage devices.

The polarization of undulator radiation can be controlled and varied in so-called APPLE (Advanced Planar Polarized Light Emitter) undulators. APPLE undulators have undergone successive refinements (reflected by their generational names of APPLE-I, II, III and X), based on the flexibility of movements of their magnet arrays; here, we consider only the basic principles.

Instead of the two Halbach magnet arrays found in 'normal' undulators discussed thus far, APPLE undulators consist of four arrays (two above, two below) that can be longitudinally shifted relative to each other [A1–A4, Fig. 1.15a] [9]. Each array consists of a periodic repetition of four different orientations [down (blue $\otimes$); reverse (green arrow); up (yellow $\odot$); and forward (red arrow), see Fig. 1.15b]. With all array components aligned longitudinally, normal linear horizontal (LH) radiation as found in 'normal' undulators is produced. Antisymmetric shifts (that is, one array moving in the positive z-direction, the other in the negative) of A1 and A4 between $-\lambda_u/2$ and $+\lambda_u/2$ will maintain linearly polarized radiation but will vary the tilt angle from linear vertical (LV at $-\lambda_u/2$), through LH (zero shift) back to LV ($+\lambda_u/2$). Conversely, symmetric movements of A1 and A4 by approximately $\pm\lambda_u/4$ will produce right circularly polarized light (at approximately $-\lambda_u/4$), or left circularly polarized light (at $+\lambda_u/4$). In this last instance, the exact shift in the arrays depends on the gap size and details of the magnet strengths. Symmetric shifts smaller than this result in elliptical polarization, that is, radiation with both a linear and circularly polarized component.

Synchrotron sources provide fluxes and brilliances many orders of magnitude greater than those from laboratory-based sources. Third-generation facilities were defined by their use of insertion devices, most importantly undulators. The fourth generation of synchrotrons, just beginning to come online at the time of writing, increase brilliances by two orders of magnitude by reducing the horizontal (orbital

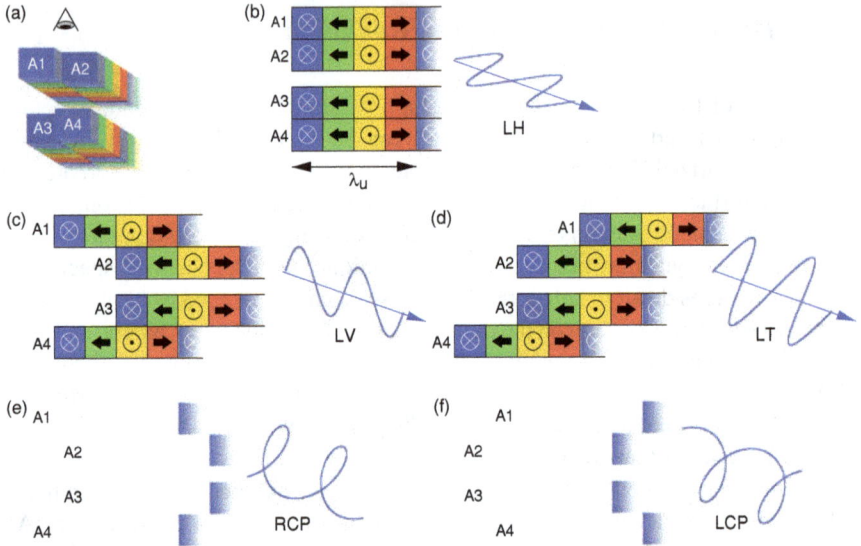

Fig. 1.15 APPLE undulators. **a** APPLE undulators consist of four magnet arrays A1, A2, A3 and A4. **b** The magnet periodicity λ_u is composed of four magnet orientations. Viewed from above, these are down (blue ⊗); reverse horizontal (green arrow); up (yellow ⊙); and forward horizontal (red arrow). When all four arrays are aligned longitudinally, linear horizontal polarization is produced. **c** By shifting arrays A1 and A4 either symmetrically or antisymmetrically by $\pm\lambda_u/2$, linear vertically polarized radiation is produced. **d** Linear polarization of any desired tilt angle can be selected by moving A1 and A4 antisymmetrically relative to A2 and A3. **e** and **f** Circularly polarized radiation is generated by symmetric shifts of A1 and A4 by approximately $\lambda_u/4$, the exact shift depending on the gap size and details of the magnetic field strengths.

plane) emittance through the implementation of MBAs. We discuss this in more detail now.

1.4 Diffraction-Limited Storage Rings

As we have already intimated, the main limit to the emittance in storage rings is due to the electron spread induced at bending magnet achromats as a consequence of quantum excitation (Sect. 1.2.6). In our discussion of DBAs in Sect. 1.3.1, the lower limit to the electron emittance, its so-called 'natural emittance', was given by (1.23). However, if we extend an arc section simply by increasing the number of dipoles from 2 to M in a so-called MBA, and *keeping the swept angle per dipole constant*, the ratio between the natural emittances of the DBA and MBA is

$$\frac{\epsilon_{x,\text{DBA}}}{\epsilon_{x,\text{MBA}}} = 3\frac{M-1}{M+1} . \tag{1.34}$$

Such an MBA, subtending an arc of angle $M/2$ times larger than that of the DBA, will have a superior emittance up to a factor of three for large M. So, for example, a 10° DBA will have a natural emittance 2.25 times that of a 35° 7BA ($M = 7$) for a fixed swept angle per dipole.

This is clearly not the way to proceed, as an increase from two bends to M will reduce the number of straights around the ring by a factor of $M/2$. A far more effective way to improve the natural emittance of an MBA can be achieved by reducing the swept angle θ per dipole. This is due to the cubic dependence of the former on the latter [see (1.23)]. So, if in the example above, the 7BA is designed to sweep the same *total angle* as the DBA, the reduction in emittance will now be $(4/9) \times (2/7)^3 = 32/3087 \approx 1/100$, thus producing emittances which assume values of the order of 100 pm rad. This is only an order of magnitude larger than the diffraction-limited photon emittance for hard X-rays of approximately 10 pm rad. Indeed, soft X-ray beamlines, for which $\lambda/4\pi \sim \epsilon_{x,\mathrm{MBA}}$, are already diffraction-limited when using MBAs.

Note also that the natural emittance scales with the square of the storage ring energy $\mathcal{E}$. The two highest-energy third-generation storage rings, namely the APS (7 GeV) and SPring8 (8 GeV) are both planning to decrease $\mathcal{E}$ to 6 GeV in their upgrades, despite the associated decrease in the highest accessible photon energies.

In most instances, because an achromat (be it a DBA or MBA) requires about the same length of 'real estate' within the storage ring, the ring's circumference C is approximately inversely proportional to θ—large rings have bend achromats that subtend smaller angles than those at small rings. This is why the large APS facility has 36 sectors, while the smaller ALS in Berkeley only has 12. Thus, for a fixed cell design, a convenient figure of merit

$$ M = \frac{\epsilon_x C^3}{\mathcal{E}^2} \tag{1.35} $$

describes how well optimized the magnet lattice is [10], as this takes into account the available real estate and electron beam energy and weighs the emittance by this, accordingly.

The newest (fourth) generation synchrotron facilities are thus dubbed 'diffraction-limited storage-rings' (DLSRs) [10–13]. A light source is said to be diffraction-limited if the horizontal electron beam emittance is smaller than that of the radiated photon beam ($\lambda/4\pi$). In practical terms, the diffraction-limited photon energy E_{DL} with a wavelength equal to $4\pi\epsilon_x$, is given by

$$ E_{\mathrm{DL}}[\mathrm{keV}] = \frac{98.66}{\epsilon_x[\mathrm{pm\,rad}]} . \tag{1.36} $$

By this metric, even a second-generation facility would be a DLSR for a beamline generating infrared radiation; in contrast, most third-generation facilities are diffraction-limited in the very soft X-ray regime at around 25 eV. Note that beamlines that use photon energies that lie near to or below the diffraction limit for the facility where they are installed gain little or nothing from the emittance of the storage ring being further improved (i.e. reduced by more sophisticated electron optics). This is the true meaning of the diffraction limit: the fundamental lower value set by the photon beam rather than the electron beam.

DLSRs are thus defined by their use of MBAs. They gain most at large-circumference storage rings, at medium storage ring energies and indeed the two greenfield DLSR facilities to date (MAX-IV and Sirius) plus the first major upgrade from third to fourth generation (ESRF-EBS) reflect these characteristics. This seemingly obvious and long-understood approach to improve the horizontal emittance via the use of MBAs and thereby the brilliance has only recently been pursued because, until now, the costs (both in hardware and real estate) and the introduction of potential mechanical misalignments associated with increasing the number of elements in the magnet lattice, were considered unacceptable. Miniaturization of these magnet lattice components and the development of multifunctional magnets machined from a single yoke block have decreased both costs and the necessary circumference of the ring to accommodate the MBAs and thus achieve these goals. The PETRA-IV project is especially interesting, as the circumference is 2.3 km, allowing exceptionally shallow angles for the arcs, promising diffraction-limited photon energies well into the hard X-ray regime, at approximately 6 keV.

Another serious obstacle to reducing the size and separation of the magnets is that the lateral dimensions (that is, the cross-section) of the storage ring vacuum vessels containing the electron beam need to become so small that pumping them with traditional pumping equipment, in particular, ion-getter pumps, becomes increasingly difficult. In recent years, however, a novel approach to achieve ultrahigh vacuum (UHV) conditions has been developed, namely the use of non-evaporable getter (NEG) coating of the inner walls of the storage ring vacuum vessels [14, 15]. NEGs are porous alloys of Al, Ti, Fe, V and Zr sintered onto the inner walls of the vacuum vessel to a thickness of the order of a micrometre, or even as little as 100–200 nm. After installation, the vacuum vessel is pumped to a moderately high vacuum using traditional pumps and then heated out to temperatures below 200 °C. This activates the NEG material, allowing pressures to drop to approximately 10^{-10} mbar in the UHV regime. Importantly, even very narrow spaces can be readily coated. Moreover, recent developments in computer numerical control of machining storage ring components allow micrometre accuracy such that geometrically near-perfect miniature vacuum vessels with cross-sections of the order of a square centimetre can be constructed.

With the advent of DLSRs and improvements in magnetic materials, undulator spectra have undergone important transformations. Figure 1.16 compares the brilliance of the same undulator at a third-generation source and at a DLSR having a 40 times smaller total horizontal emittance. In addition to the expected 40 times increase in peak brilliance, the DLSR-spectrum is substantially cleaner: the lobes seen on the low-energy flanks of the spectral peaks for the third-generation source are completely absent in the DLSR spectrum. This is because the horizontal width of the electron beam is much smaller (typically by an order of magnitude). In third-generation facilities, the horizontal electron beam width is, at approximately 100 μm, two orders of magnitude larger than the oscillation amplitude A, which is of the order

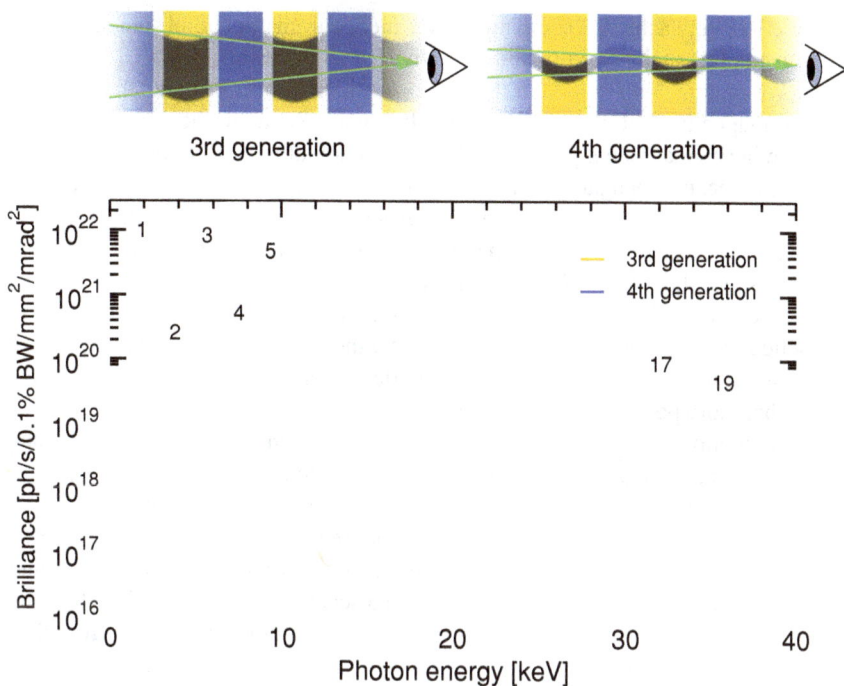

Fig. 1.16 Comparison of the brilliance of undulator spectra at third- and fourth-generation facilities. Top: the lateral extent of the electron beam passing through an undulator at third-generation facilities is approximately two orders of magnitude larger than the oscillation amplitude A, while at DLSRs, it might only be approximately $10A$. Bottom: as a consequence, an observer on-axis will 'see' less off-axis radiation, given by (1.33), thereby suppressing the lobes on the low-energy flanks of the main spectral maxima. Note also the enhanced brilliance at the spectral peaks for the DLSR. Both simulated spectra were generated for a U12 undulator (that is, $\lambda_u = 12$ mm) containing 120 magnet periods, for $K = 1.6$, 400 mA and a storage ring energy of 2.4 GeV. Courtesy Marco Calvi, Paul Scherrer Institute.

of a micrometre. This means that an observer on axis at the DLSR-undulator will see a much smaller contribution from emission from off-axis electrons, and it is these that produce the low-energy lobes in the spectra, according to (1.33). Many experiments do not require the relative BWs of the order of 10^{-4} provided by crystal monochromators and would profit from using the entire flux from any given harmonic. The relative BW of the mth undulator harmonic is $1/mN$, where N is the number of undulator periods. For the lower harmonics, this is of the order of 0.005–0.01; for higher harmonics, the relative BW does not drop as steeply as $1/m$, as gradually, the off-axis contributions do begin to leak into the peaks, causing them to broaden marginally. Nonetheless, the spectral quality remains sufficiently high to use the entire flux of any given harmonic for small-period undulators at DLSRs. Note, however, spectral filtering is still required to remove the other harmonics. This can be effectively achieved using multilayer monochromators or, in some instances, refractive optics such as compound refractive lenses or prisms [16].

Improvements don't stop here, however. A critical aspect of undulator design is that the magnetic field must be exceedingly homogeneous in the x-direction (that is, in the horizontal plane, perpendicular to the undulator axis) in the region where the electron beam propagates. This has meant that, for third-generation facilities, the magnet yokes need to be a few centimetres wide or more in the x-direction. The approximately five to ten times smaller extent of electron beams passing through undulators installed in DLSRs means that the yokes can be of the order of one centimetre. This allows the design of compact magnetic 'funnels', concentrating magnetic field lines and thus increasing the magnetic field strength, which, in turn, allows for a more compact design and shorter undulator periods. Moreover, this means that more periods (N) can be fit into a given undulator length L.

The reduction of the horizontal electron emittance in DLSRs thus lends many potential scientific opportunities and will drive exciting innovations along the full technology chain, from the sources (in particular, undulators), through improved X-ray optics (regarding minimization of optical imperfections and aberrations), detector technology, to handling of large data volumes. The first DLSR, MAX-IV in Lund, Sweden, came online in Summer 2016, while a second greenfield facility. Sirius in Campinas, Brazil and the ESRF, the first facility to undergo an upgrade from third generation to DLSR status, both began pilot experiments in early 2020.

Several orders of magnitude higher peak brilliance are provided by high-gain X-ray free-electron lasers (XFELs). The machine science and technologies for this paradigm shift in X-ray sources are based on fundamentally different principles, and consequently, XFELs are discussed separately in the next section.

1.5   X-Ray Free-Electron Lasers

Radiation from fourth-generation DLSRs has some important common features with laser radiation: it is very intense, collimated and, in the case of radiation from undulators, partially monochromatic. An important distinction, however, is that the degree of transverse (spatial) coherence of synchrotron radiation, although much improved at DLSRs compared to that produced by third-generation facilities, is still only of the order of a few percent in the hard X-ray regime. This is because, although radiation from any single electron travelling along an undulator is coherent, there is no spatial (i.e. phase) correlation between different electrons, and hence their combined output remains largely incoherent. In contrast, visible lasers are normally close to being 100% coherent.

Moreover, the shortest pulse duration of X-rays from synchrotrons is a few tens of picoseconds and may be as large as a few hundred picoseconds in the case of DLSRs. Lasers in the visible and near-visible regimes can have pulse lengths as small as 70 attoseconds in some exceptional cases; more representatively, pulse lengths of a few femtoseconds are routinely achieved. The production of femtosecond light pulses has been motivated by studies of the dynamics of chemical reactions—this is one of the most important goals in science, as a biomolecular understanding of the processes of life, plus most industrial chemical processes, depend intimately on understanding transient intermediate states. Indeed, the holy grail of physical chemistry is the direct observation of the making and breaking of chemical bonds and following reactions through the reconfiguration of the atomic structure. In order to achieve this, one needs a probe that can spatially resolve the atomic positions, measured in angstroms, thus necessitating the use of hard X-rays. Moreover, in order to record a movie of chemical processes, each 'frame' must be shorter than the speed of the process divided by the desired resolution, which we have just stated should be of the order of an angstrom. Typical velocities associated with atomic motion are of the order of $1000 \, \mathrm{m \, s^{-1}}$, hence each movie frame should be separated by no more than approximately 100 fs. For example, phonons and molecular rotations have time scales of approximately 100–1000 fs, while molecular vibrations, crucial to the process of chemical reactions, are shown to have periodicities of the order of 10–100 fs.

X-ray free-electron lasers (XFELs) were developed to fulfil this goal. They provide extremely intense and short pulses of X-rays. XFELs and synchrotron facilities differ substantially in this respect (see Table 1.1). The total average number of photons per second delivered to a third- or fourth-generation synchrotron beamline after monochromatization (typically between 10^{13} and $10^{14} \, \mathrm{s^{-1}}$) is comparable to that of an XFEL beamline, although the latter can vary by orders of magnitude, depending on the facility repetition rate and whether the beam is monochromatized or not. In contrast, their time structures are very different: synchrotrons deliver pulses with the same frequency as the RF supply, typically several hundred million pulses per second ($1/\Delta t$); each pulse is maybe a few tens of picosecond long ($\Delta \tau$) and contains the order of 10^4–10^5 photons. XFELs deliver anything between approximately 100 and several hundreds of thousands of pulses per second, depending on the electron gun source and the properties of the accelerating LINAC. Most importantly, each XFEL pulse is only a few tens of femtoseconds long (and can be even shorter, down to a femtosecond) and contains the order of 10^{12} photons. The peak arrival rate of photons at XFELs is thus approximately 10 billion times higher than that at synchrotron beamlines.

XFELs thus provide the tools needed to study the dynamics, physical properties and structure of materials with unsurpassed spatial detail and on a time scale over a thousand times shorter than that which is possible at synchrotron facilities.

The beam quality in storage rings is limited by the stochastic competing processes described above in radiation equilibrium. High-gain XFELs are made possible by a runaway process called self-amplified spontaneous emission (SASE). Because, in contrast to synchrotrons, the electrons in XFELs are not stored and require only a few microseconds to traverse the length of the entire XFEL facility, they are far less perturbed than are the equilibrated electrons in a storage ring, meaning they can maintain a very low emittance defined by the electron gun, acceleration and bunch compression mechanism.

Table 1.1 Comparison of orders of magnitude synchrotron and XFEL radiation[†] properties. [†]XFEL values derive from LCLS unless otherwise stated. 8-keV photons assumed. *EuroXFEL time structure: 2700 pulses at 4.5 MHz, 10 such bursts per second. **photons s^{-1} 0.1% BW^{-1}. [‡]After Si(111) monochromator, $\Delta\nu/\nu = 1.4 \times 10^{-4}$. [¶]Unmonochromatized, full SASE spectrum. [§]23 kW during pulse burst

Property	Synchrotron	XFEL
$\Delta\tau$	50 – 400 ps	1 – 100 fs
Δt	5 ns	$10^{-2} - 2 \times 10^{-7}$ s*
Average flux**	2×10^{14}	10^{14}
Peak flux**	6×10^{15}	2×10^{25}
#$h\nu$/pulse	4×10^{4} [‡]	4×10^{12} [¶]
Peak power	1 W [‡]	10^{11} W[¶]
Average power	25 mW[‡]	600 mW[¶] – 140 W[¶]*[§]

The linear architecture of XFELs is very different than that of synchrotron facilities; it is summarized in the following.

1.5.1   XFEL Architecture

The fate of electrons in a high-gain XFEL proceeds as follows. In order to generate pulses of electrons, a ps laser is focussed on to the surface of a photoemitter. The resulting ps-duration electron pulse contains a charge of the order of 300 pC (or 2×10^9 electrons) and is accelerated in a first LINAC to energies of the order of several hundred MeV (see Fig. 1.17). Shorter laser pulse durations containing the same total number of electrons are excluded due to Coulomb repulsion, which would blow up the beam size and thus spoil the emittance. However, once the electrons become relativistic after acceleration in the first, short, LINAC, the bunch can be compressed without spoiling the emittance. This is achieved in one or more bunch compressors; thereafter, the electrons are further accelerated in a longer LINAC, before entering a very long undulator array. It is while the electrons are travelling along the undulator that the SASE process takes place, resulting in the emission of femtosecond-duration pulses of X-rays.

The initial few-ps-duration electron bunch produced by the low-emittance gun has a transient peak current of about 50 A (300 pC/6 ps). Bunch compression is required in order to reduce the pulse duration to approximately 300 fs and produce the several thousand amperes peak current required to induce SASE (Fig. 1.18).

The necessary factor of approximately 20–50 increase is realized by adjusting the RF phase in the first accelerator module (LINAC 1 in Fig. 1.17) to allow the electrons to 'surf' down the slope of the sinusoidal RF field—the leading electrons experience a slightly smaller acceleration by the sinusoidal electric field of the RF cavity and are thus accelerated less than those electrons towards the back of the

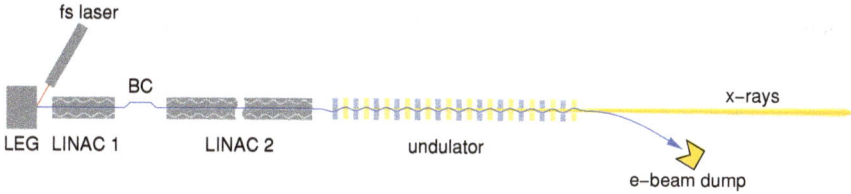

Fig. 1.17 Schematic of XFEL architecture. Electron bunches are emitted from a low-emittance gun (LEG) irradiated by picosecond laser pulses. They are then accelerated in a short LINAC (LINAC 1), then compressed using one or more bunch compressor magnet chicanes (BC); they are then further accelerated using a much longer LINAC (LINAC 2) before entering a long undulator, typically a few hundred metres in length. The SASE process along the undulator produces extremely intense X-ray pulses with durations of the order of 50 fs.

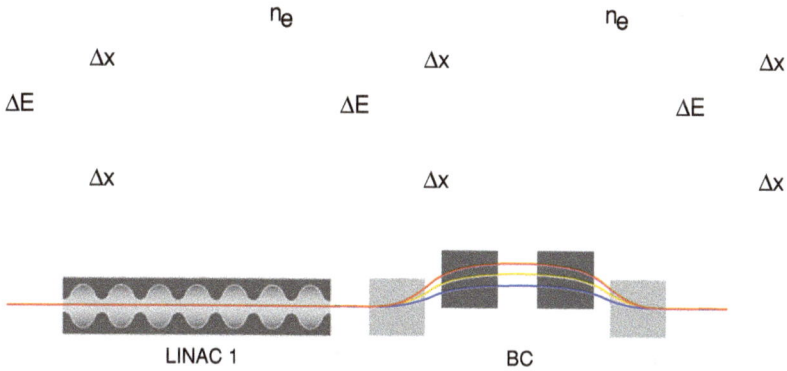

Fig. 1.18 Bunch compression in high-gain XFELs. Before entering the first LINAC, the electron density n_e and spread in kinetic energy ΔE are both relatively low. The phase of the LINAC RF-field relative to the passage of the bunch is so adjusted to induce a larger spread of the electrons' kinetic energies, whereby the bunch's trailing edge is made to be more energetic than the leading edge. Compression is achieved by allowing the bunch to pass through a four-dipole chicane, where the faster electrons at the back can catch up with the slower electrons at the front, thanks to the shorter path that they execute.

bunch (a phenomenon referred to as 'chirping'). However, the electrons' velocities are already so close to the speed of light that any differences are far too small to allow the faster electrons at the back of the bunch to catch up with the slower electrons positioned further forward and thereby squeeze the bunch length. Instead, the bunch passes through a magnetic four-dipole chicane. The trailing (high-energy) electrons execute a shorter path through the chicane because they are less deviated by the chicane's magnetic field. This shorter path means that they catch up with the less energetic leading electrons, which are more deviated by the magnet chicane. This compression thus shortens the bunch duration to approximately 300 fs (equating to a bunch length of 100 μm) and increases n_e by a factor of approximately 20–50. The increased energy spread of the chirped electron beam after bunch compression must be removed through 'dechirping', which can be achieved in one of several ways [18, 19].

1.5.2 The SASE Process

As the electrons begin to propagate down the undulator, they initially emit X-rays independently and stochastically and are bathed in this radiation. They will interact with this EM field. We begin our discussion of SASE by determining the magnitude of these forces in conventional undulators at synchrotrons.

Classical electromagnetism tells us that the power transmitted per unit area by an EM plane wave is given by

$$\frac{P}{A} = \frac{E_0 H_0}{2}, \tag{1.37}$$

where A is the cross-section of the beam, E_0 is the amplitude of the electric field component and H_0 is the amplitude of the magnetic intensity, which, in vacuum, is related to the magnetic field strength amplitude B_0 by

$$H_0 = \frac{B_0}{\mu_0}, \tag{1.38}$$

where $\mu_0 = 4\pi \times 10^{-7}$ m kg C^{-2} is the permeability of free space (given here in SI units). Also from classical electromagnetic theory, it emerges that

$$B_0 = \frac{E_0}{c}. \tag{1.39}$$

From (1.37) to (1.39), we thus obtain an areal power density

$$\frac{P}{A} = \frac{E_0^2}{2\mu_0 c}. \tag{1.40}$$

A typical synchrotron undulator may have a source size of $A = 100 \times 10$ μm^2, produce light bunches of 50-ps duration, each containing 5×10^6 1-Å photons (within the full width of a harmonic). This equates to an areal power density of the order of 1.6×10^{11} W m^{-2}. From (1.39) and (1.40), one calculates an electric field amplitude $E_0 \approx 10^7$ V m^{-1} and $B_0 \approx 40$ mT. This latter value is nearly two orders of magnitude smaller than that imposed by the undulator's magnet array. We can thus conclude that, in a conventional third- or fourth-generation synchrotron facility, the forces of the electromagnetic field generated by undulator radiation on the electrons' trajectory are negligible.

In the case of XFELs, however, the beam is tailored to have as low an emittance and as high an electron density as possible, through the low-emittance gun and bunch compression, respectively. The motivation for this is exactly to induce forces through the generated radiation that are sufficiently large that they do indeed have an impact on the energy and spatial distribution of the electron bunch. Although at the upstream end of the XFEL undulator, this interaction is still very weak, it is sufficient to seed the runaway process of SASE, which is now briefly described.

The ratio of the Lorentz force and electric field force experienced by an electron bathed in EM radiation is

$$F_L = eE\frac{v}{c} = -F_E\beta \ . \tag{1.41}$$

The electrons within XFELs are highly relativistic and hence to a high degree of accuracy, $\beta \approx 1$ and $F_L \approx -F_E$ (the fractional imbalance in the forces is equal to $\Delta F/F_E = 1/2\gamma^2 \sim 2 \times 10^{-9}$). Thus, for an electron moving exactly along the axis of the EM radiation (that is, ignoring for the time being the oscillations induced by the magnet array of an undulator), the electric and Lorentz forces are equal and opposite and point perpendicularly to the beam propagation.

If the electrons move in a straight line parallel to the EM plane wave, both the electric and magnetic forces (which anyway cancel each other out) always act at right angles to the electrons' direction of propagation and thus could transfer energy to or from the electrons. What happens if we now allow the electrons to move in directions that are not precisely parallel to the EM radiation, such as in the slalom path induced by an undulator's magnet array? Let us now consider those positions exactly in between the magnet poles, where the electrons have their maximal transverse velocity.

The first observation to make is that, because the magnetic field of the emitted EM radiation always lies perpendicular to the plane of the electrons' trajectory, it cannot transfer energy, but only vary the direction of the electrons' motion. In contrast, the electric field part of the emitted radiation has a component parallel to the electrons' trajectory and can either decelerate them or accelerate them. Consequently, some electrons will be accelerated while others are decelerated. These two opposing interactions cause the electrons to form microbunches within the 'normal' bunch, separated by a distance equal to the wavelength of the light they both generate and are bathed in. Although, for hard X-rays, the 'normal' bunch length of approximately 100 μm equates to approximately 10^6 microbunches, SASE only begins in the coincidentally most intense portion of the bunch. As such, SASE is a stochastic process. The duration of the SASE radiation therefore depends on the degree of bunch compression and the integrated charge of the initial bunch as it enters the undulator array. Low-charge bunches will thus produce XFEL pulses with lower peak brilliance but with durations that can be shorter than 1 fs. More commonly, 'standard' bunch durations are a few tens of femtoseconds.

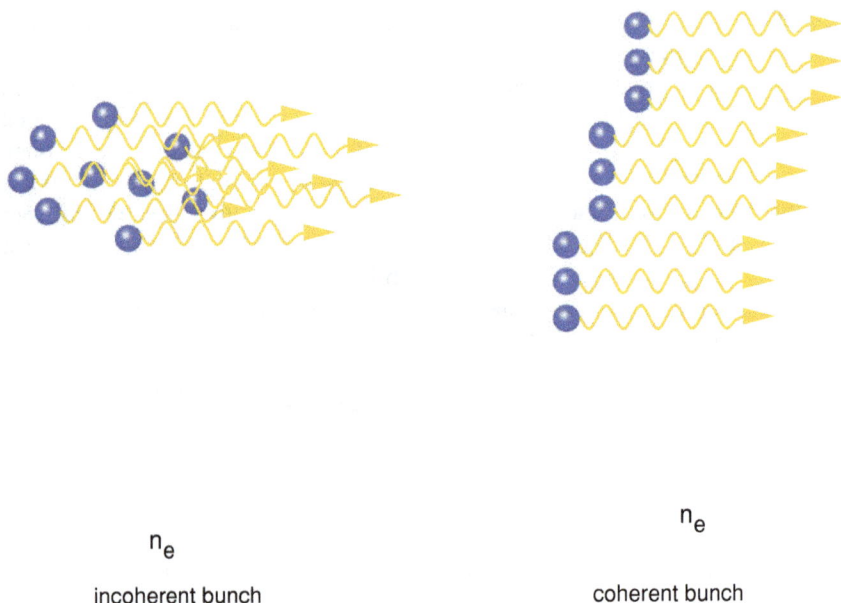

n_e

incoherent bunch

n_e

coherent bunch

Fig. 1.19 Incoherent and coherent radiation from an ensemble (bunch) of electrons. If the relative longitudinal positions of the electrons bear no correlation to the radiation they emit, the relative phases of that radiation are likewise random. From normal statistics, it can be shown that their summed amplitude is equal to the amplitude of the radiation emitted from one electron multiplied by the square root of the number of electrons (n_e) in the bunch. The intensity (proportional to the square of the amplitude) is therefore proportional to n_e. In contrast, the radiation induced by SASE is fully coherent, as the emission from all electrons is in phase. The amplitudes of the radiation from each of the n_e electrons thus add linearly, resulting in an increase in intensity by n_e^2.

SASE is a runaway effect. The more the electrons bunch together into microbunches, the stronger is the radiation they produce, leading to stronger associated forces on those electrons and still tighter microbunching.

An important property of the microbunches produced by SASE is the following. In the case of normal SR, the emission of radiation from n_e electrons within a bunch is stochastic and uncorrelated, and hence the vector addition of the individual amplitudes will follow normal statistics and, on average, be equal to $A_{tot} = \sqrt{n_e}A$, where A is the amplitude of the contribution to the radiation of any one electron. But the intensity is proportional to the square of the amplitude and is therefore proportional to n_e.

In the case of XFEL SASE radiation, the initial bunch is modified so that it becomes a set of microbunches separated from their neighbours by the wavelength of light produced by the undulator interference. It follows that the individual contributions to the radiation amplitude will all be more or less in phase and the electrons will all emit coherently. Thus, the amplitudes add linearly as they all have the same phase and hence $A_{tot} = n_e A$. The intensity will be proportional to n_e^2 (see Fig. 1.19). This is the key to the 'runaway' amplification in SASE: the more the electrons bunch and radiate in phase (that is, coherently), the stronger is the EM-field interaction with them. This in turn further enhances the microbunching phenomenon associated with coherent emission, thus increasing the degree of coherent emission in a positive feedback loop.

This runaway instability causes the light intensity to grow exponentially along the undulator until the process saturates [20]. The microbunches produced by the SASE process each contain approximately $n_e = 10^9$ electrons, resulting in a peak brilliance for XFELs approximately one billion times larger than that from fourth-generation synchrotron sources!

The quality of a high-gain XFEL is encapsulated in the dimensionless so-called Pierce, or FEL, parameter, ρ_{FEL}, given approximately by

$$\rho_{FEL} = \left[\frac{\lambda_u^2 r_0 n_e K^2}{32\pi \gamma^3} \right]^{1/3} , \qquad (1.42)$$

where λ_u is the undulator period length, $r_0 = 2.82 \times 10^{-15}$ m is the Thomson scattering length (also known as the classical electron radius), N_e is the electron density of the pulse bunch as it enters the undulator, K is the deviation parameter, and $\gamma = \mathcal{E}/m_e c^2$ is the Lorentz factor. Equation (1.42) can be re-expressed in practical units as

$$\rho_{FEL} = \frac{1.55 \times 10^{-5}}{\mathcal{E}[GeV]} \left[(\lambda_u[mm])^2 N_e[mm^{-3}]K^2 \right]^{1/3} . \qquad (1.43)$$

Electron densities in the compressed electron bunch entering the undulator are typically $10^4 \, \mu m^{-3}$ (that is, 10^9 electrons in a volume of the order of $10^5 \, \mu m^3$); the undulator period $\lambda_u = 15$–30 mm, and K-values of approximately 2–3 are normal. The LINACs produce electron energies of the order of 10 GeV. Because of the weak cube-root dependency of ρ_{FEL} on all the design variables except $\mathcal{E}$, its spread is fairly narrow across all facilities designed or commissioned to date, and normally assumes values of around 5×10^{-4}.

ρ_{FEL} determines three important properties of high-gain XFEL radiation. Firstly, it describes the fraction of the electrons' power converted at SASE saturation to photon power, that is

$$\rho_{FEL} = \frac{P_{ph}}{P_e} . \qquad (1.44)$$

The larger is ρ_{FEL}, the more efficiently electron energy is converted into photon energy. ρ_{FEL} also describes the relative spectral BW at saturation,

$$\rho_{FEL} = \frac{\Delta\omega}{\omega} , \qquad (1.45)$$

which means that at photon energies of the order of 10 keV, the instantaneous BW (that is, the BW of any one FEL pulse) is of the order of 5 eV. This is over an order of magnitude larger than the transmitted BW after monochromatization with a diamond (004) single crystal. In addition, the stochastic nature of the initial spontaneous generation of SASE radiation is responsible for the central position of the SASE spectrum jittering from pulse to pulse by a few tens of eV. This is acceptable for many types of experiment, such as in serial femtosecond crystallography, but can be problematic in others, especially in spectroscopy. The bandwidth can be reduced in several ways including self-seeding, or the use of chicanes in so-called high-brightness SASE. It lies outside the remit of this overview to detail these here [3].

In contrast to the transverse coherence length, the longitudinal coherence length of SASE pulses [which depends inversely on the BW, see (1.14)] is small, of the order of 200 nm in the hard X-ray regime.

Thirdly, the distance L_G within an XFEL undulator required to obtain a gain in SASE radiation by a factor $e \approx 2.72$ is inversely proportional to ρ_{FEL} and is given by

$$L_G = \frac{\lambda_u}{4\pi\sqrt{3}\rho_{FEL}}. \tag{1.46}$$

Inserting our known values for λ_u and ρ_{FEL}, we obtain values for L_G of the order of a few metres. Saturation of SASE occurs after a gain of approximately five orders of magnitude. But $10^5 \approx e^{11}$, and hence the total undulator length should exceed L_G by a factor of 11 or more. The LCLS hard X-ray undulator is over 110 m long.

Hard XFEL radiation can therefore directly track atomic motions in condensed matter and vibrations with periods of the order of picoseconds down to tens of femtoseconds, a hitherto inconceivable scientific endeavour. XFELs deliver peak brilliances many orders of magnitude greater than radiation from third- and fourth-generation synchrotron sources, 100% transverse coherent radiation, and pulse durations typically of a few tens of femtoseconds, but which can also be tailored to be less than one femtosecond [21, 22].

1.5.3   Concluding Remarks

The main objective of this section was to summarize the machine physics of XFELs in a manner that is accessible to a wide spectrum of potential users. XFEL technology and science are bound to further develop and morph in the next decades, as, in contrast to synchrotron science and technology, XFELs are still very much in their infancy. Nonetheless, they are maturing rapidly, thanks in no small part to the knowledge base already established for synchrotron facilities. Because the radiation produced by XFELs differs so greatly from SR in the peak brilliance and associated very short pulse durations, experimental methods commonly used at synchrotrons need more often than not to be entirely rethought in order to be carried out at XFELs; indeed, some experimental methods at synchrotrons are wholly excluded at XFELs.

On the other hand, some experimental techniques first developed at XFELs, in particular serial femtosecond crystallography, are now being adopted and adapted for synchrotron light. XFELs should thus not be considered as superior alternatives to synchrotrons, but more as complementary facilities.

Many experimental disciplines, particularly in the fields fs-time resolved studies and nonlinear light-matter interactions, have emerged using XFELs and are still rapidly evolving; this is the reason, in fact, why XFELs were developed in the first place. It will be intriguing to see how these develop and expand in the forthcoming years and decades. The future of photon science indeed looks bright!

1.6 Summary

This chapter has introduced the reader to the basic physics of radiation produced by accelerated electrons and the recent advances in machine technologies which have resulted in the development of DLSRs and XFELs. The number of dedicated X-ray sources has burgeoned over the last three decades and has opened the field of synchrotron science, transforming it into a multidisciplinary enterprise, nowadays attracting some hundred thousand scientists across the broadest spectrum of the natural sciences.

References

1. E. Dagotto, Complexity in strongly correlated electronic systems. Science **309**, 257 (2005). https://doi.org/10.1126/science.1107559
2. A. Ohtomo, H.Y. Hwang, A high-mobility electron gas at the $LaAlO_3/SrTiO_3$ heterointerface. Nature **427**, 423 (2004). [Corrigendum: Nature **441**, 120 (2006).] https://doi.org/10.1038/nature02308 [https://doi.org/10.1038/nature04773]
3. P.R. Willmott, *An Introduction to Synchrotron Radiation: Techniques and Applications*, 2nd edn. (Wiley, Hoboken, 2019). https://doi.org/10.1002/9781119280453
4. R. Chasman, G.K. Green, E.M. Rowe, Preliminary design of a dedicated synchrotron radiation facility. IEEE Trans. Nucl. Sci. **22**, 1765 (1975). https://doi.org/10.1109/TNS.1975.4327987
5. A. Wolski, Low-emittance storage rings, in *CERN Accelerator School Series – Advanced Accelerator Physics Course*, Trondheim 2013, ed. by W. Herr (CERN, Geneva, 2014), p. 245. https://doi.org/10.5170/CERN-2014-009.245
6. K. Halbach, Physical and optical properties of rare earth cobalt magnets. Nucl. Instrum. Methods Phys. Res. **187**, 109 (1981). https://doi.org/10.1016/0029-554X(81)90477-8
7. K. Halbach, J. Chin, E. Hoyer, H. Winick, R. Cronin, J. Yang, Y. Zambre, A permanent magnet undulator for SPEAR. IEEE Trans. Nucl. Sci. **28**, 3136 (1981). https://doi.org/10.1109/TNS.1981.4332031
8. P.R. Willmott, D. Meister, S.J. Leake, M. Lange, A. Bergamaschi, M. Böge, M. Calvi, C. Cancellieri, N. Casati, A. Cervellino, Q. Chen, C. David, U. Flechsig, F. Gozzo, B. Henrich, S. Jäggi-Spielmann, B. Jakob, I. Kalichava, P. Karvinen, J. Krempasky, A. Lüdeke, R. Luscher, S. Maag, C. Quitmann, M.L. Reinle-Schmitt, T. Schmidt, B. Schmitt, A. Streun, I. Vartiainen, M. Vitins, X. Wang, R. Wullschleger, The materials science beamline upgrade at the swiss light source. J. Synchrotron Radiat. **20**, 667 (2013). https://doi.org/10.1107/S0909049513018475

9. M. Calvi, C. Camenzuli, E. Prat, T. Schmidt, Transverse gradient in apple-type undulators. J. Synchrotron Radiat. **24**, 600 (2017). https://doi.org/10.1107/S1600577517004726

10. M. Borland, Progress toward an ultimate storage ring light source. J. Phys. Conf. Ser. **425**, 042016 (2013). https://doi.org/10.1088/1742-6596/425/4/042016

11. M. Eriksson, Å. Andersson, S. Biedron, M. Demirkan, G. LeBlanc, L.J. Lindgren, L. Malmgren, H. Tarawneh, E. Wallén, S. Werin, MAX 4, a 3 GeV light source with a flexible injector, in *Proceedings of the 8th European Particle Accelerator Conference (EPAC 2002)*, Paris 2002, p. 686. https://accelconf.web.cern.ch/e02/PAPERS/TUPLE005.pdf

12. M. Eriksson, L.J. Lindgren, M. Sjöström, E. Wallén, L. Rivkin, A. Streun, Some small-emittance light-source lattices with multi-bend achromats. Nucl. Instrum. Methods Phys. Res. Sect. A **587**, 221 (2008). https://doi.org/10.1016/j.nima.2008.01.068

13. E. Weckert, The potential of future light sources to explore the structure and function of matter. IUCrJ **2**, 230 (2015). https://doi.org/10.1107/S2052252514024269

14. C. Benvenuti, J.M. Cazeneuve, P. Chiggiato, F. Cicoira, A. Escudeiro Santana, V. Johanek, V. Ruzinov, J. Fraxedas, A novel route to extreme vacua: the non-evaporable getter thin film coatings. Vacuum **53**, 219 (1999). https://doi.org/10.1016/S0042-207X(98)00377-7

15. C. Benvenuti, P. Chiggiato, P. Costa Pinto, A. Escudeiro Santana, T. Hedley, A. Mongelluzzo, V. Ruzinov, I. Wevers, Vacuum properties of TiZrV non-evaporable getter films. Vacuum **60**, 57 (2001). https://doi.org/10.1016/S0042-207X(00)00246-3

16. I. Inoue, T. Osaka, K. Tamasaku, H. Ohashi, H. Yamazaki, S. Goto, M. Yabashi, An X-ray harmonic separator for next-generation synchrotron X-ray sources and X-ray free-electron lasers. J. Synchrotron Radiat. **25**, 346 (2018). https://doi.org/10.1107/S160057751800108X

17. A.H. Zewail, Femtochemistry: atomic-scale dynamics of the chemical bond. J. Chem. Phys. A **104**, 5660 (2000). https://doi.org/10.1021/jp001460h

18. S. Antipov, S. Baturin, C. Jing, M. Fedurin, A. Kanareykin, C. Swinson, P. Schoessow, W. Gai, A. Zholents, Experimental demonstration of energy-chirp compensation by a tunable dielectric-based structure. Phys. Rev. Lett. **112**, 114801 (2014). https://doi.org/10.1103/PhysRevLett.112.114801

19. P. Emma, M. Venturini, K.L.F. Bane, G. Stupakov, H.S. Kang, M.S. Chae, J. Hong, C.K. Min, H. Yang, T. Ha, W.W. Lee, C.D. Park, S.J. Park, I.S. Ko, Experimental demonstration of energy-chirp control in relativistic electron bunches using a corrugated pipe. Phys. Rev. Lett. **112**, 034801 (2014). https://doi.org/10.1103/PhysRevLett.112.034801

20. P. Emma, R. Akre, J. Arthur, R. Bionta, C. Bostedt, J. Bozek, A. Brachmann, P. Bucksbaum, R. Coffee, F.-J. Decker, Y. Ding, D. Dowell, S. Edstrom, A. Fisher, J. Frisch, S. Gilevich, J. Hastings, G. Hays, Ph. Hering, Z. Huang, R. Iverson, H. Loos, M. Messerschmidt, A. Miahnahri, S. Moeller, H.-D. Nuhn, G. Pile, D. Ratner, J. Rzepiela, D. Schultz, T. Smith, P. Stefan, H. Tompkins, J. Turner, J. Welch, W. White, J. Wu, G. Yocky, J. Galayda, First lasing and operation of an ångstrom-wavelength free-electron laser. Nat. Photonics **4**, 641 (2010). https://doi.org/10.1038/nphoton.2010.176

21. J. Feldhaus, J. Arthur, J.B. Hastings, X-ray free-electron lasers. J. Phys. B: At. Mol. Opt. Phys. **38**, S799 (2005). https://doi.org/10.1088/0953-4075/38/9/023

22. G. Margaritondo, P. Rebernik Ribič, A simplified description of X-ray free-electron lasers. J. Synchrotron Radiat. **18**, 101 (2011). https://doi.org/10.1107/S090904951004896X

2

Magnetism and its General Concepts

Stephen J. Blundell

Abstract I review some general concepts in magnetism including the nature of magnetic exchange (direct, indirect and superexchange), and how exchange interactions play out in multiple spin systems. The nature of atomic orbitals and the way in which they interact with the spin system is also considered. Several examples are also treated, including the Jahn–Teller interaction and its role in the properties in layered manganites.

2.1 Introduction

Magnetic properties are found in a wide variety of materials. In order to explain magnetism we need to consider a range of different behaviours in many different types of magnetic system. Consider the following: Fe and Ni are both metallic elements and exhibit ferromagnetism; MnO is an insulating oxide with a three-dimensional antiferromagnetic structure; La_2CuO_4 is a layered material which exhibits antiferromagnetism but, when doped, becomes superconducting; some compounds do not order magnetically but show frustrated effects with an abundance of slow dynamics; some molecules become single-molecule magnets in which the individual molecules show quantum tunnelling of magnetization and a range of other interesting properties. Theories of magnetism have to explain all these materials and more.

For a start, we must realize that a classical approach will not work. The Bohr–van Leeuwen theorem [1] states that in a classical system there is no thermal equilibrium magnetization. We can prove this in outline as follows: in classical statistical mechanics the partition function Z for N particles, each with charge q, is proportional to

$$\int \int \cdots \int \exp\left[-\beta E(\{r_i, p_i\})\right] \, dr_1 \cdots dr_N \, dp_1 \cdots dp_N , \qquad (2.1)$$

S. J. Blundell (✉)
Clarendon Laboratory, Oxford University, Department of Physics, Parks Road,
Oxford OX1 3PU, UK
e-mail: stephen.blundell@physics.ox.ac.uk

where $\beta = 1/k_B T$, k_B is the Boltzmann constant, T is the temperature and $i = 1, \ldots, N$. Here $E(\{r_i, p_i\})$ is the energy associated with the N charged particles having positions $r_1, r_2, \ldots, r_N$, and momenta $p_1, p_2, \ldots, p_N$. The integral is, therefore, over a $6N$-dimensional phase space ($3N$ position coordinates, $3N$ momentum coordinates). The effect of a magnetic field is to shift the momentum of each particle by an amount qA. We must, therefore, replace p_i by $p_i - qA$. The limits of the momentum integrals go from $-\infty$ to $+\infty$ so this shift can be absorbed by shifting the origin of the momentum integrations. Hence the partition function is not a function of magnetic field, and so neither is the free energy $F = -k_B T \log Z$. Thus the magnetic moment $m = -(\partial F/\partial B)_T$ must be zero in a classical system.

Thus, we need quantum mechanics to make further progress. In this chapter, I will not provide an exhaustive review of magnetism (fuller treatments can be found elsewhere, e.g. [2–5]) but focus on a few key issues and some selected examples. To begin our discussion, it is helpful to note the energy scales inherent in magnetic problems. First, there is the kinetic energy which is on the eV scale. This typically takes a value like $\hbar^2 \pi^2/(2mL^2)$, where L is a length scale and this expression is the familiar one for particle in a box. This is an energy cost and arises because it takes energy to put an electron in a small box. Second comes the potential energy which is also on a similar scale and takes a form such as $e^2/(4\pi \epsilon_0 L)$. This will be a negative energy if considering the attraction between an electron and a nucleus (and becomes larger and more negative as L decreases) and positive if considering electron–electron repulsion. Atoms are the size they are because of a compromise between kinetic energy wanting the atom to be infinite size and the potential energy wanting the atom to be zero size. Because one energy goes as L^{-2} and the other as $-L^{-1}$ a compromise can be reached (and this is essentially the derivation of the Bohr radius). Both the kinetic and potential energies are large and are typically $\gg k_B T$. Next, we have to add the spin–orbit interaction which is typically much smaller, usually in the meV, and the magnetocrystalline anisotropy, which in cubic materials is in the μeV. These effects will turn out to be very important in magnetic materials, but they are small perturbations to the main interactions and will mainly come into play only once the magnetic order is established by the dominant interactions.

2.2 Exchange

The exchange interaction arises from the kinetic and potential energy in bonds between atoms. To see how this comes about, we begin by recalling simple results for the molecular orbitals in H_2 [see Fig. 2.1a]. We label the two hydrogen atoms A and B and write the wave function $|\psi\rangle$ as a linear combination of atomic orbitals $|\psi_A\rangle$ and $|\psi_B\rangle$ so that

$$|\psi\rangle = c_A|\psi_A\rangle + c_B|\psi_B\rangle . \tag{2.2}$$

Fig. 2.1 a Two hydrogen atoms A and B can lower their energy by forming a hydrogen molecule H_2. **b** The bonding and antibonding molecular orbitals σ and σ^*

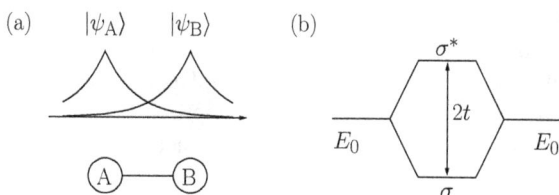

The Hamiltonian can be written as a sum of the kinetic energy and two terms for the potential energy due to the attraction to each hydrogen nucleus so that

$$\hat{\mathcal{H}} = -\frac{\hbar^2}{2m}\nabla^2 + V_A + V_B \ . \tag{2.3}$$

We then need to solve the equation $\hat{\mathcal{H}}|\psi\rangle = E|\psi\rangle$. The diagonal integral E_0, which can be approximated by the binding energy of the electron at one of the centres for a hydrogen *atom*, is given by

$$E_0 = \langle\psi_A|\hat{\mathcal{H}}|\psi_A\rangle \ . \tag{2.4}$$

The transfer integral (also known as the hopping integral or resonance integral) t is given by

$$t = \langle\psi_A|\hat{\mathcal{H}}|\psi_B\rangle \ . \tag{2.5}$$

In the simplest approximation (the Hückel approximation), the overlap integrals are given by $S_{ij} = \langle\psi_i|\psi_j\rangle = \delta_{ij}$ and hence the secular equation $|H_{ij} - E S_{ij}| = 0$ can be written as

$$\begin{vmatrix} E_0 - E & t \\ t & E_0 - E \end{vmatrix} = 0 \ , \tag{2.6}$$

and hence

$$E = E_0 \pm t \ . \tag{2.7}$$

The eigenfunctions for these solutions are the symmetric solution

$$|\sigma\rangle = \frac{|\psi_A\rangle + |\psi_B\rangle}{\sqrt{2}} \tag{2.8}$$

which costs energy $E_0 - t$ and the antisymmetric solution

$$|\sigma^*\rangle = \frac{|\psi_A\rangle - |\psi_B\rangle}{\sqrt{2}} \tag{2.9}$$

which costs energy $E_0 + t$. These are known as the bonding and antibonding states, respectively [see Fig. 2.1b]. The hydrogen molecule has two electrons so the σ level is full and the σ^* level is empty, thus saving the energy overall and leading to the H_2 molecule being a stable entity. The molecule He_2 does not form because it has four electrons and would, therefore, involve filling both σ and σ^* and thus saves no energy (and in fact, outside the Hückel approximation, it turns out that $\sigma^* - E_0 > E_0 - \sigma$ and so helium bonding costs more energy than it saves).

2.2.1 Direct Exchange

Exchange interactions are nothing more than a consequence of electrostatic interactions and the familiar interplay between potential energy and kinetic energy that we see in chemical bonds. Consider a simple model with just two electrons which have spatial coordinates r_1 and r_2, respectively. The wave function for the joint state can be written as a product of single electron states, so that if the first electron is in state $\psi_a(r_1)$ and the second electron is in state $\psi_b(r_2)$, then the joint wave function is $\psi_a(r_1)\psi_b(r_2)$. However, this product state does not obey exchange symmetry, since if we exchange the two electrons we get $\psi_a(r_2)\psi_b(r_1)$, which is not a multiple of what we started with. Therefore, the only states which we are allowed to make are symmetrized or antisymmetrized product states which behave properly under the operation of particle exchange.

For electrons, the overall wave function must be antisymmetric so the spin part of the wave function must either be an antisymmetric singlet state χ_S ($S = 0$) in the case of a symmetric spatial state or a symmetric triplet state χ_T ($S = 1$) in the case of an antisymmetric spatial state. Therefore, we can write the wave function for the singlet case Ψ_S and the triplet case Ψ_T as

$$\Psi_S = \frac{1}{\sqrt{2}}[\psi_a(r_1)\psi_b(r_2) + \psi_a(r_2)\psi_b(r_1)]\,\chi_S$$

$$\Psi_T = \frac{1}{\sqrt{2}}[\psi_a(r_1)\psi_b(r_2) - \psi_a(r_2)\psi_b(r_1)]\,\chi_T\,, \qquad (2.10)$$

where both the spatial and spin parts of the wave function are included. The energies of the two possible states are

$$E_S = \int \Psi_S^* \hat{\mathcal{H}} \Psi_S \, dr_1 \, dr_2$$

$$E_T = \int \Psi_T^* \hat{\mathcal{H}} \Psi_T \, dr_1 \, dr_2\,,$$

with the assumption that the spin parts of the wave function χ_S and χ_T are normalized. The difference between the two energies is

$$E_S - E_T = 2\int \psi_a^*(r_1)\psi_b^*(r_2)\hat{\mathcal{H}}\psi_a(r_2)\psi_b(r_1)\, dr_1 \, dr_2\,. \qquad (2.11)$$

For a singlet state $S_1 \cdot S_2 = -\frac{3}{4}$ while for a triplet state $S_1 \cdot S_2 = \frac{1}{4}$. Hence the Hamiltonian can be written in the form of an 'effective Hamiltonian'

$$\hat{\mathcal{H}} = \frac{1}{4}(E_S + 3E_T) - (E_S - E_T)S_1 \cdot S_2 . \tag{2.12}$$

This is the sum of a constant term and a term which depends on spin. The constant can be absorbed into other constant energy terms, but the second term is more interesting. The exchange constant J is defined by

$$J = \frac{E_S - E_T}{2} = \int \psi_a^*(r_1)\psi_b^*(r_2)\hat{\mathcal{H}}\psi_a(r_2)\psi_b(r_1)\, dr_1\, dr_2 , \tag{2.13}$$

and hence the spin-dependent term in the effective Hamiltonian can be written as

$$\hat{\mathcal{H}}^{\text{spin}} = -2J S_1 \cdot S_2 . \tag{2.14}$$

If $J > 0$, $E_S > E_T$ and the triplet state $S = 1$ is favoured. If $J < 0$, $E_S < E_T$ and the singlet state $S = 0$ is favoured. Thus, the exchange interaction compares two different configurations that are tied to the singlet and triplet spin states, but the energy difference associated with exchange comes from the difference in those two configurations worked out from an integral [see (2.13)] over the spatial coordinates. Thus the spins are really there just to label the two different spatial states and are inextricably tied to the spatial wave functions by the Pauli principle; the exchange interaction is *really* between spatial wave functions, even though we tend to think about it as between the spin parts that really just come along for the ride!

Equation (2.14) is relatively simple to derive for two electrons, but generalizing to a many-body system is far from trivial. It motivates the Hamiltonian of the Heisenberg model:

$$\hat{\mathcal{H}} = -\sum_{ij} J_{ij} S_i \cdot S_j , \tag{2.15}$$

where J_{ij} is the exchange constant between the ith and jth spins. The factor of 2 is omitted because the summation includes each pair of spins twice. Another way of writing (2.15) is

$$\hat{\mathcal{H}} = -2\sum_{i>j} J_{ij} S_i \cdot S_j , \tag{2.16}$$

where the $i > j$ avoids the 'double-counting' and hence the factor of two returns. It is worth noting that there are different conventions for the definition of J that are in use in the literature. I call these the J-convention and the $2J$-convention and they are summarized in Fig. 2.2. Note that it is also possible to choose the sign of J so that $J > 0$ means ferromagnetic (as here) or antiferromagnetic. Both choices are found in the literature.

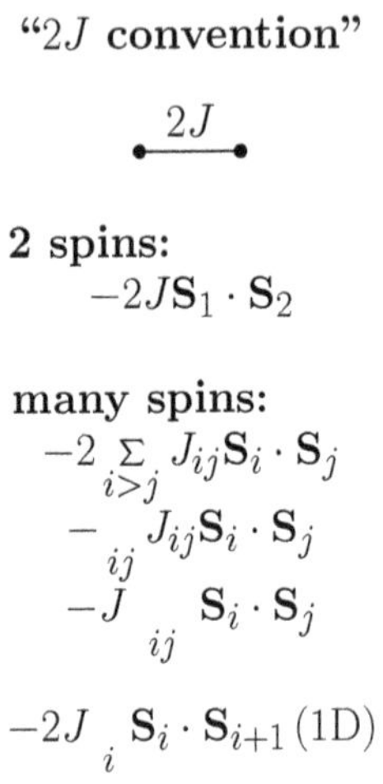

Fig. 2.2 The two different conventions used for the definition of J. In this chapter (and in [2]), we are using the $2J$ convention (so that $2J$ is the energy associated with a single pairwise interaction between two spins). The various alternative expressions that one can use for the Heisenberg interaction are shown under the heading 'many spins', as well as an expression for a one-dimensional (1D) chain of spins

"$2J$ convention"

$$\underset{\bullet \qquad \bullet}{2J}$$

2 spins:
$$-2J\mathbf{S}_1 \cdot \mathbf{S}_2$$

many spins:
$$-2 \sum_{i>j} J_{ij}\mathbf{S}_i \cdot \mathbf{S}_j$$
$$-\sum_{ij} J_{ij}\mathbf{S}_i \cdot \mathbf{S}_j$$
$$-J \sum_{ij} \mathbf{S}_i \cdot \mathbf{S}_j$$

$$-2J \sum_i \mathbf{S}_i \cdot \mathbf{S}_{i+1} \,(1\mathrm{D})$$

2.2.2 Indirect Exchange

If the electrons on neighbouring magnetic atoms interact via an exchange interaction, this is known as direct exchange. This is because the exchange interaction proceeds directly without the need for an intermediary, and this was considered in the previous section.

Very often direct exchange cannot be an important mechanism in controlling the magnetic properties because there is insufficient direct overlap between neighbouring magnetic orbitals. For example, in rare earths the $4f$ electrons are strongly localized and lie very close to the nucleus, with little probability density extending significantly further than about a tenth of the interatomic spacing. This means that the direct exchange interaction is unlikely to be very effective in rare earths. Even in transition metals, such as Fe, Co and Ni, where the $3d$ orbitals extend further from the nucleus, it is extremely difficult to justify why direct exchange should lead to the observed magnetic properties. These materials are metals which means that the role of the conduction electrons should not be neglected, and a correct description needs to be taken into account of both the localized and band character of the electrons.

In metals the exchange interaction between magnetic ions can be mediated by the conduction electrons. A localized magnetic moment spin-polarizes the conduction electrons and this polarization in turn couples to a neighbouring localized magnetic moment a distance r away. The exchange interaction is thus indirect because it does not involve direct coupling between magnetic moments. It is known as the RKKY interaction (or also as itinerant exchange). The name RKKY is used because of the initial letters of the surnames of the discoverers of the effect, Ruderman, Kittel, Kasuya and Yosida [6–8]. The coupling takes the form of an r-dependent exchange interaction $J_{\mathrm{RKKY}}(r)$ given by

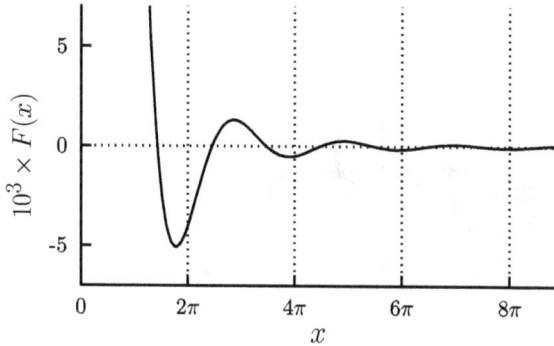

Fig. 2.3 The real space susceptibility of a free electron gas is given by $\chi(r) = 2k_F^3\mu_0\mu_B^2 g(E_F)F(2k_Fr)/\pi$ where $F(x) = (-x\cos x + \sin x)/x^4$ is the function illustrated. A localized spin in a free electron gas, therefore, gives rise to an effective exchange $J_{RKKY} \propto F(2k_Fr)$ and this is proportional to $\cos(2k_Fr)/r^3$ when $r \gg k_F^{-1}$

$$J_{RKKY}(r) \propto \frac{\cos(2k_Fr)}{r^3} , \tag{2.17}$$

at large r (assuming a spherical Fermi surface of radius k_F). The interaction is long range and has an oscillatory dependence on the distance between the magnetic moments (see Fig. 2.3). Hence depending on the separation it may be either ferromagnetic or antiferromagnetic. The coupling is oscillatory with wavelength π/k_F because of the sharpness of the Fermi surface.

2.2.3 Superexchange

A number of ionic solids, including some oxides and fluorides, have magnetic ground states. For example, MnO [see Fig. 2.4a] and MnF_2 are both antiferromagnets, though this observation appears at first sight rather because there is no direct overlap between the electrons on Mn^{2+} ions in each system. The exchange interaction is normally very short ranged so that the longer ranged interaction that is operating in this case must be in some sense 'super' (think of Superman leaping over buildings, a skill not afforded to ordinary mortals).

The origin of superexchange is the possibility of mixing in excited states to lower the energy. The favouring of antiferromagnetic superexchange in a linear Mn–O–Mn bond arises from the fact that the excited states are allowed, while for the ferromagnetic arrangement these excited states are forbidden [see Fig. 2.4b]. One can consider this problem with a toy model based on a Hubbard-style Hamiltonian (see, e.g. [9]) which may be written as

$$\hat{\mathcal{H}} = -t\sum_{\langle ij\rangle\sigma}\hat{c}_{i\sigma}^\dagger\hat{c}_{j\sigma} + U\sum_i\hat{n}_{i\uparrow}\hat{n}_{i\downarrow} , \tag{2.18}$$

(a)

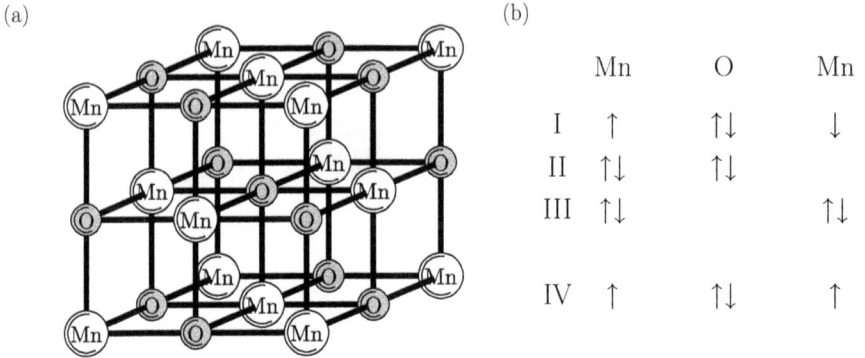

(b)

	Mn	O	Mn
I	↑	↑↓	↓
II	↑↓	↑↓	
III	↑↓		↑↓
IV	↑	↑↓	↑

Fig. 2.4 a The crystal structure of MnO. Nearest neighbour pairs of Mn^{2+} (manganese) ions are connected via O^{2-} (oxygen) ions. **b** A simple model of superexchange for a Mn–O–Mn bond. I: the antiferromagnetic ground state with opposite spins on the two Mn ions and a pair of electrons on the oxygen anion. II and III: two excited states of the antiferromagnetic ground state in which the electrons from (I) hop back and forth. IV: the competing ferromagnetic ground state. This is energetically more costly because the excited states analogous to (II) and (III) are not available because of the Pauli exclusion principle

where the first sum is over nearest neighbours; thus energy is lowered by hopping (the first term on the right) but there is an energy penalty for double occupancy (the second term on the right) due to the Coulomb repulsion energy U. Let us now restrict this model to a system with two possible sites for electrons (here we are ignoring the intermediate oxygen to for simplicity). We can start by putting a single electron with spin ↑ into the system. Using a basis $|\uparrow, 0\rangle$ and $|0, \uparrow\rangle$, the Hamiltonian is given by

$$\hat{\mathcal{H}} = \begin{pmatrix} 0 & -t \\ -t & 0 \end{pmatrix}, \tag{2.19}$$

because with one electron there is no possibility of a Coulomb penalty, and so the only energy to worry about is the energy saving you get from hopping. This is the same as the H_2 problem we considered earlier and the eigenvalues are $\pm t$ and so the lowest energy state is the bonding state, just as before.

Now let us put a second electron into the system with opposite spin to the first. Now using a basis such that a general state can be written as

$$|\psi\rangle = a|\uparrow\downarrow, 0\rangle + b|\uparrow, \downarrow\rangle + c|\downarrow, \uparrow\rangle + d|0, \uparrow\downarrow\rangle, \tag{2.20}$$

and we can easily show that in this basis the Hubbard Hamiltonian is

$$\hat{\mathcal{H}} = \begin{pmatrix} U & t & -t & 0 \\ t & 0 & 0 & t \\ -t & 0 & 0 & -t \\ 0 & t & -t & U \end{pmatrix}, \tag{2.21}$$

where the minus signs appear because of the exchange symmetry. The eigenvalues are 0, U and $(U/2) \pm \sqrt{(U/2)^2 + 2t^2}$, so in the limit that $t/U \ll 1$ the last pair of eigenvalues are $U + 2t^2/U + O(t^4/U^3)$ and $-2t^2/U + O(t^4/U^3)$. Thus, the ground state has energy $-2t^2/U$. If we try the same problem again with two electrons with the *same* spin then they cannot sit on the same site because of the Pauli exclusion principle. Thus the only state possible is $|\uparrow, \uparrow\rangle$ and this has energy $E = 0$. Thus, there is an energy saving in having the two electrons with opposite spin because you can go lower than $E = 0$ and have $E = -2t^2/U + O(t^4/U^3)$. This means that the exchange interaction has a magnitude $J \approx 2t^2/U$. The moral of the story is that by having the possibility to mix in the higher energy states in which two spins sit on the same site (costing U), it is possible to lower the overall energy. The antiferromagnetic arrangement allows this process to happen; the ferromagnetic arrangement forbids it. Superexchange can be considered in more detail [10] and can in certain circumstances be ferromagnetic. The size and sign of the superexchange interaction is codified in the Goodenough–Kanamori–Anderson rules [11–14].

2.3 Consequences of the Heisenberg Exchange Interaction

We have seen that at the heart of the exchange interaction is a term $\hat{S}_a \cdot \hat{S}_b$, a simple scalar product between two spin operators. If that scalar product is expanded, we have

$$\hat{S}_a \cdot \hat{S}_b = \hat{S}_a^x \hat{S}_b^x + \hat{S}_a^y \hat{S}_b^y + \hat{S}_a^z \hat{S}_b^z = \hat{S}_a^z \hat{S}_b^z + \frac{1}{2}(\hat{S}_a^+ \hat{S}_b^- + \hat{S}_a^- \hat{S}_b^+) , \qquad (2.22)$$

where the raising and lowering operators $\hat{S}^+$ and $\hat{S}^-$ are defined by

$$\hat{S}^+ = \hat{S}^x + i\hat{S}^y$$
$$\hat{S}^- = \hat{S}^x - i\hat{S}^y . \qquad (2.23)$$

Although the term $\hat{S}_a^z \hat{S}_b^z$ in (2.22) seems to be simple enough to handle, the term $\frac{1}{2}(\hat{S}_a^+ \hat{S}_b^- + \hat{S}_a^- \hat{S}_b^+)$ will give rise to flip-flop processes in which simultaneously an up-spin labelled a is lowered and a down-spin labelled b is raised, or vice versa. This part of the interaction has profound effects.

2.3.1 Two Interacting Spin-$\frac{1}{2}$ Particles

In this section, we will consider two spin-$\frac{1}{2}$ particles coupled by a scalar interaction described by a Hamiltonian $\hat{\mathcal{H}} = A\hat{S}_a \cdot \hat{S}_b$, where $\hat{S}_a$ and $\hat{S}_b$ are the operators for the spins for the two particles. We can also write the total spin operator $\hat{S}^{\text{tot}} = \hat{S}_a + \hat{S}_b$ so that

Table 2.1 The eigenstates of $\hat{S}_a \cdot \hat{S}_b$ for a two-spin system and the corresponding values of m_s, s and the eigenvalue of $\hat{S}_a \cdot \hat{S}_b$

Eigenstate	m_s	s	$\hat{S}_a \cdot \hat{S}_b$
$\lvert\uparrow\uparrow\rangle$	1	1	$\frac{1}{4}$
$\dfrac{\lvert\uparrow\downarrow\rangle + \lvert\downarrow\uparrow\rangle}{\sqrt{2}}$	0	1	$\frac{1}{4}$
$\lvert\downarrow\downarrow\rangle$	-1	1	$\frac{1}{4}$
$\dfrac{\lvert\uparrow\downarrow\rangle - \lvert\downarrow\uparrow\rangle}{\sqrt{2}}$	0	0	$-\frac{3}{4}$

$$(\hat{S}^{\text{tot}})^2 = (\hat{S}_a)^2 + (\hat{S}_b)^2 + 2\hat{S}_a \cdot \hat{S}_b .\tag{2.24}$$

In quantum mechanics, when you combine the angular momentum of two spin-$\frac{1}{2}$ particles you have the 'addition law' that $\frac{1}{2} + \frac{1}{2} = 0, 1$. You can think of this simply as arising from the fact that you can combine the two moments together constructively or destructively. Alternatively, imagine adding two classical vectors J_1 and J_2 together but varying the angle between them. In that case, the resulting vector $J_1 + J_2$ would have length ranging from $|J_1 - J_2|$ to $J_1 + J_2$ (where $J_1 = |J_1|$ and $J_2 = |J_2|$). More formally, combining the representations of two spin-$\frac{1}{2}$ together yields a representation

$$D^{(\frac{1}{2})} \otimes D^{(\frac{1}{2})} = D^{(0)} \oplus D^{(1)} .\tag{2.25}$$

In other words, the result of combining two spin-$\frac{1}{2}$ particles is a combined object with spin quantum number $s = 0$ or 1. The eigenvalue of $(\hat{S}^{\text{tot}})^2$ is $s(s + 1)$ which is therefore either 0 or 2 for the cases of $s = 0$ or 1. The eigenvalues of both $(\hat{S}_a)^2$ and $(\hat{S}_b)^2$ are $\frac{3}{4}$. Hence from (2.24)

$$\hat{S}_a \cdot \hat{S}_b = \begin{cases} +\frac{1}{4} & \text{if } s = 1 \\ -\frac{3}{4} & \text{if } s = 0 . \end{cases}\tag{2.26}$$

The system, therefore, has two energy levels for $s = 0$ and 1 with energies given by

$$E = \begin{cases} +\frac{A}{4} & \text{if } s = 1 \\ -\frac{3A}{4} & \text{if } s = 0 . \end{cases}\tag{2.27}$$

The degeneracy of each state is $2s + 1$, so that the $s = 0$ state is a singlet (a single energy level) and the $s = 1$ state is a triplet (three energy levels). The z component of the spin of this state, m_s, can only equal 0 for the singlet, but can be $-1, 0,$ or 1 for the triplet. Thus the product of two spin-$\frac{1}{2}$ representations, which have a dimensionality of $2 \times 2 = 4$, gives rise to states $s = 0$ (singlet) and $s = 1$ (triplet), which have a total dimensionality of $1 + 3 = 4$.

We have considered the eigenvalues of $\hat{S}_a \cdot \hat{S}_b$, but what about the eigenstates? The most straightforward basis to consider is

$$|\uparrow\uparrow\rangle, \quad |\uparrow\downarrow\rangle, \quad |\downarrow\uparrow\rangle, \quad |\downarrow\downarrow\rangle. \tag{2.28}$$

The first arrow refers to the z component of the spin labelled a and the second arrow refers to the z component of the spin labelled b. The eigenstates of $\hat{S}_a \cdot \hat{S}_b$ are linear combinations of these basis states and are listed in Table 2.1. The value of m_s is equal to the sum of the z components of the individual spins. Since the eigenstates are a mixture of states in the original basis, we cannot know both the z components of the original spins and the total spin of the resultant entity. This is a general feature which will become more important in more complicated situations.

The basis in (2.28) also fails to satisfy the condition that the overall wave function must be antisymmetric with respect to exchange of the two electrons. Since the wave function is a product of a spatial function $\psi_{\text{space}}(\boldsymbol{r}_1, \boldsymbol{r}_2)$ and the spin function χ, the spatial wave function can be either symmetric or antisymmetric with respect to exchange of electrons. For example, the spatial wave function

$$\psi_{\text{space}}(\boldsymbol{r}_1, \boldsymbol{r}_2) = \frac{\phi(\boldsymbol{r}_1)\xi(\boldsymbol{r}_2) \pm \phi(\boldsymbol{r}_2)\xi(\boldsymbol{r}_1)}{\sqrt{2}} \tag{2.29}$$

is symmetric $(+)$ or antisymmetric $(-)$ with respect to exchange of electrons depending on the $\pm$. This type of symmetry is known as exchange symmetry. In (2.29), $\phi(\boldsymbol{r}_i)$ and $\xi(\boldsymbol{r}_i)$ are single-particle wave functions for the ith electron. Whatever is the exchange symmetry of the spatial wave function, the spin-wave function χ must have the opposite exchange symmetry. Hence χ must be antisymmetric when the spatial wave function is symmetric and vice versa. This is in order that the product $\psi_{\text{space}}(\boldsymbol{r}_1, \boldsymbol{r}_2) \times \chi$ is antisymmetric overall.

States such as $|\uparrow\uparrow\rangle$ and $|\downarrow\downarrow\rangle$ are clearly symmetric under exchange of electrons, but exchanging the two electrons in $|\uparrow\downarrow\rangle$ yields $|\downarrow\uparrow\rangle$ which is not a multiple of $|\uparrow\downarrow\rangle$. Thus $|\uparrow\downarrow\rangle$, and also by an identical argument $|\downarrow\uparrow\rangle$, are both neither symmetric nor antisymmetric under exchange of the two electrons. The true eigenstates must, therefore, be linear combinations of these two states (see Table 2.1). The state $(|\uparrow\downarrow\rangle + |\downarrow\uparrow\rangle)/\sqrt{2}$ is symmetric under exchange of electrons (in common with the other two $s = 1$ states) while the state $(|\uparrow\downarrow\rangle - |\downarrow\uparrow\rangle)/\sqrt{2}$ (the $s = 0$ state) is antisymmetric under exchange of electrons.

The energy levels are shown in Fig. 2.5. Without the flip-flop term $\frac{1}{2}(\hat{S}_a^+ \hat{S}_b^- + \hat{S}_a^- \hat{S}_b^+)$ in (2.22), the Hamiltonian is simply $\hat{\mathcal{H}} = A\hat{S}_a^z \hat{S}_b^z$ and this leads to two degenerate doublets as shown. The upper doublet (consisting of the states $|\uparrow\uparrow\rangle$ and $|\downarrow\downarrow\rangle$) is unchanged when the flip-flop terms are switched on. The lower doublet (consisting of the states $|\uparrow\downarrow\rangle$ and $|\downarrow\uparrow\rangle$, although strictly speaking it should of course be the symmetric and antisymmetric combinations of these two states) splits with the addition of the flip-flop terms to make $\hat{\mathcal{H}} = A\hat{S}_a \cdot \hat{S}_b$ and the symmetric combination

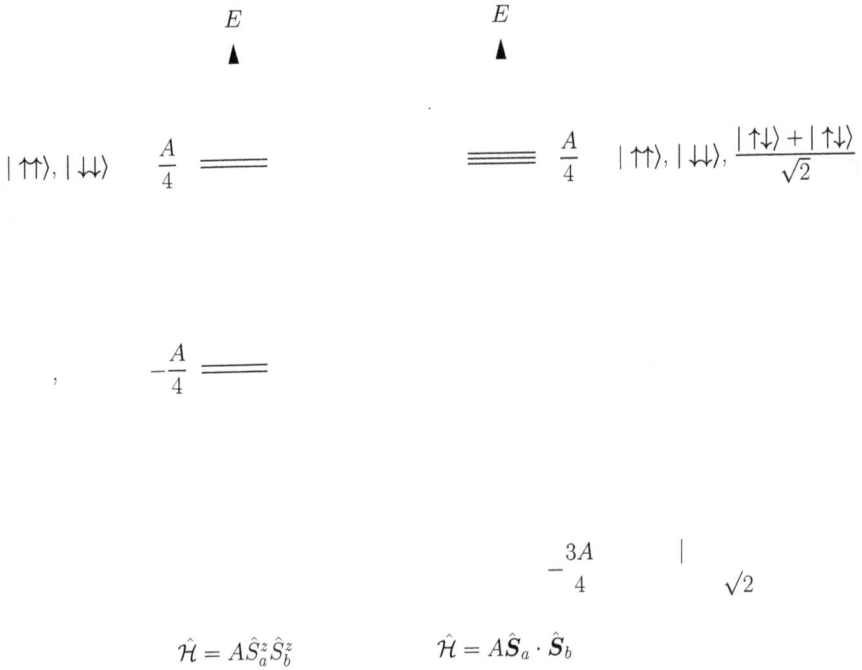

Fig. 2.5 The energy levels for the two Hamiltonians: $\hat{\mathcal{H}} = A\hat{S}_a^z \hat{S}_b^z$ and $\hat{\mathcal{H}} = A\hat{\boldsymbol{S}}_a \cdot \hat{\boldsymbol{S}}_b$

rises up in energy to $A/4$ and the antisymmetric combination is lowered to $-3A/4$, becoming the ground state when $A > 0$.

2.3.2 A Chain of Spins

Let us now consider not just two spins but a one-dimensional ferromagnetic chain of spin-$\frac{1}{2}$ moments described by the Heisenberg model. In one dimension, the Hamiltonian for the Heisenberg model can be written as

$$\hat{\mathcal{H}} = -2J \sum_i \left[\hat{S}_i^z \hat{S}_{i+1}^z + \frac{1}{2}(\hat{S}_i^+ \hat{S}_{i+1}^- + \hat{S}_i^- \hat{S}_{i+1}^+) \right], \qquad (2.30)$$

where $J > 0$. The ground state Φ consists of all spins aligned [see Fig. 2.6a] and this is an eigenstate of $\hat{\mathcal{H}}$ so that $\hat{\mathcal{H}}|\Phi\rangle = -NS^2 J|\Phi\rangle$. Now to create an excitation, let us flip a spin at site j, so let us now consider a state

$$|j\rangle = \hat{S}_j^-|\Phi\rangle \qquad (2.31)$$

(a)
$$|\Phi\rangle = |\cdots \uparrow\uparrow\uparrow\uparrow\uparrow\uparrow\uparrow\uparrow\uparrow \cdots\rangle$$

$$|j\rangle = |\cdots \uparrow\uparrow\uparrow\uparrow\downarrow\uparrow\uparrow\uparrow \cdots\rangle$$
$$ j$$

$$|q\rangle = \frac{1}{\sqrt{N}} \sum_j e^{iqR_j}|j\rangle$$

(b)

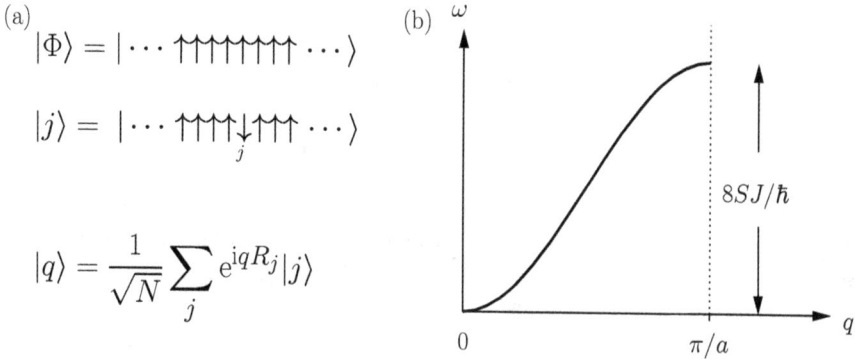

Fig. 2.6 **a** The ground state of a one-dimensional ferromagnet in the Heisenberg model is $|\Phi\rangle$. An excited state $|j\rangle$ has a single flipped spin at site j. The spin-wave state $|q\rangle$ is a delocalized spin flip. **b** The dispersion relation for the spin waves

which is simply the ground state with the spin at site j flipped [see Fig. 2.6a]. By flipping a spin, we have changed the total spin of the system by $\frac{1}{2} - (-\frac{1}{2}) = 1$. This excitation, therefore, has integer spin and is a boson. If we apply the Hamiltonian to this new state, we get

$$\hat{\mathcal{H}}|j\rangle = 2\left[(-NS^2J + 2SJ)|j\rangle - SJ|j+1\rangle - SJ|j-1\rangle\right] , \qquad (2.32)$$

which is not a constant multiplied by $|j\rangle$, so this state is not an eigenstate of the Hamiltonian. Nevertheless, we can diagonalize the Hamiltonian by looking for plane wave solutions of the form

$$|q\rangle = \frac{1}{\sqrt{N}} \sum_j e^{iqR_j}|j\rangle . \qquad (2.33)$$

The state $|q\rangle$ is essentially a flipped spin delocalized (smeared out) across all the sites [see Fig. 2.6a] and is known as a spin wave or a magnon. The state $|q\rangle$ is also an eigenstate of an operator exchanging any two spins, which is not the case for $|j\rangle$. Since $|q\rangle$ is a linear combination of states like $|j\rangle$ which represent a single flipped spin, the total spin in the z-direction of $|q\rangle$ itself has the value $NS - 1$. It is then straightforward to show that

$$\hat{\mathcal{H}}|q\rangle = E(q)|q\rangle , \qquad (2.34)$$

where

$$E(q) = -2NS^2J + 4JS(1 - \cos qa) . \qquad (2.35)$$

The energy of the excitation is then $\hbar\omega = 4JS(1 - \cos qa)$ and is plotted in Fig. 2.6b. At small q, $\hbar\omega \approx 2JSq^2a^2$, so that $\omega \propto q^2$. In three dimensions, the density of states is given by $g(q)\,dq \propto q^2\,dq$, which leads to

$$g(\omega)d\omega \propto \omega^{1/2}\,d\omega \qquad (2.36)$$

at low temperature where only small q and small ω are important. The spin waves are quantized in the same way as lattice waves. The latter are termed phonons, and so in the same way the former are termed magnons. They are bosons and have a spin of one.

The number of magnon modes excited at temperature T, n_{magnon}, is calculated by integrating the magnon density of states over all frequencies after multiplying by the Bose factor, $[\exp(\hbar\omega/k_\text{B}T) - 1]^{-1}$, which must be included because magnons are bosons. Thus the result is given by

$$n_{\text{magnon}} = \int_0^\infty \frac{g(\omega)\,d\omega}{\exp(\hbar\omega/k_\text{B}T) - 1}\,, \qquad (2.37)$$

which can be evaluated using the substitution $x = \hbar\omega/k_\text{B}T$. At low temperature, where $g(\omega) \propto \omega^{1/2}$ in three dimensions, this yields the result

$$n_{\text{magnon}} = \left(\frac{k_\text{B}T}{\hbar}\right)^{3/2} \int_0^\infty \frac{x^{1/2}\,dx}{e^x - 1} \propto T^{3/2}\,. \qquad (2.38)$$

Since each magnon mode which is thermally excited reduces the total magnetization by one (because each magnon mode is a delocalized single reversed spin), then at low temperature the reduction in the spontaneous magnetization from the $T = 0$ value is given by

$$\frac{M(0) - M(T)}{M(0)} \propto T^{3/2}\,. \qquad (2.39)$$

This result is known as the Bloch $T^{3/2}$ law. If one repeats this calculation in two dimensions (rather than three), the integral diverges, showing that magnons spontaneously form at all non-zero temperatures, thereby destroying any magnetization. The impossibility of spontaneous magnetization in two dimensions for the Heisenberg model is known as the Mermin–Wagner theorem [15–17] (see also [9]).

2.3.3   Three Spins

Let us return now to an apparently simpler system and consider three spins on the corners of an equilateral triangle. We will put the exchange interaction to be negative and thus the system is frustrated. If we put the first spin up, the next one down, then we have a dilemma of how to arrange the third one because we cannot satisfy the antiferromagnetic interactions on every bond. The solution has to be one of compromise and in fact the ground state of the classical Heisenberg model on a triangle is the so-called 120° state shown in Fig. 2.7a.

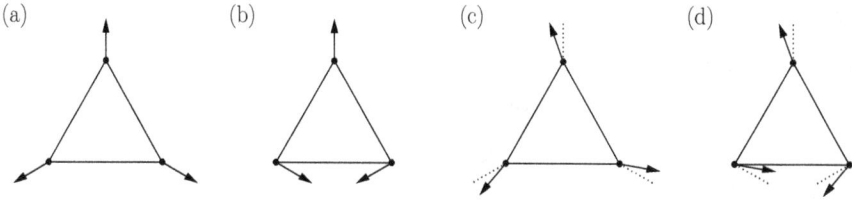

Fig. 2.7 The ground state of the classical Heisenberg model is the 120° state shown in (**a**), though a version with opposite chirality (**b**) is also possible. If each of the three spins are rotated by a constant angle [(**c**) and (**d**)] then additional ground state configurations can be obtained

In fact, there are some other possible solutions since we can choose to wind the spins round the triangle in two different ways. The configuration in Fig. 2.7b also has a 120° angle between adjacent spins but has the opposite *chirality* to that of Fig. 2.7a. Moreover, the Heisenberg model only cares about the *relative angle* between spins, not their absolute orientation, and therefore the configurations in Fig. 2.7c and d are also part of the ground state manifold.

Let us now solve the problem quantum mechanically. The law of addition of angular momentum now gives $\frac{1}{2} + \frac{1}{2} + \frac{1}{2} = \frac{1}{2}, \frac{1}{2}, \frac{3}{2}$. Now three two-dimensional representations (for three spin-$\frac{1}{2}$) have a dimensionality of $2^3 = 8$ which is equal to two two-dimensional representation and a four-dimensional representation (for two spin-$\frac{1}{2}$ and a single spin-$\frac{3}{2}$, so $2^3 = 2 + 2 + 4$). Another way of writing this combination is

$$D^{(\frac{1}{2})} \otimes D^{(\frac{1}{2})} \otimes D^{(\frac{1}{2})} = 2D^{(\frac{1}{2})} \oplus D^{(\frac{3}{2})} \,. \tag{2.40}$$

For three spins we have that

$$\hat{S}^{\text{tot}} = \hat{S}_1 + \hat{S}_2 + \hat{S}_3 \tag{2.41}$$

and hence

$$(\hat{S}^{\text{tot}})^2 = \hat{S}_1^2 + \hat{S}_2^2 + \hat{S}_3^2 + 2 \sum_{\langle i,j \rangle} \hat{S}_i \cdot \hat{S}_j \,, \tag{2.42}$$

and so using the facts that the eigenvalue of $(\hat{S}^{\text{tot}})^2$ is $S^{\text{tot}}(S^{\text{tot}} + 1)$ and the eigenvalue of $\hat{S}_i^2$ is $\frac{1}{2}(\frac{1}{2} + 1) = \frac{3}{4}$, we have

$$\sum_{\langle i,j \rangle} \hat{S}_i \cdot \hat{S}_j = \frac{1}{2} \left(S^{\text{tot}}(S^{\text{tot}} + 1) - 3 \times \frac{3}{4} \right) \,. \tag{2.43}$$

We have two cases: (i) $S^{\text{tot}} = \frac{3}{2}$ implies that $\sum_{\langle i,j \rangle} \hat{S}_i \cdot \hat{S}_j = \frac{3}{4}$; (ii) $S^{\text{tot}} = \frac{1}{2}$ implies that $\sum_{\langle i,j \rangle} \hat{S}_i \cdot \hat{S}_j = -\frac{3}{4}$. The energy levels are drawn in Fig. 2.8 and consist of two degenerate doublets at $E = -3A/4$ ($S = \frac{1}{2}$) and a quartet at $E = 3A/4$.

Fig. 2.8 The energy levels for a triangle of spins consist of two degenerate doublets and a quartet

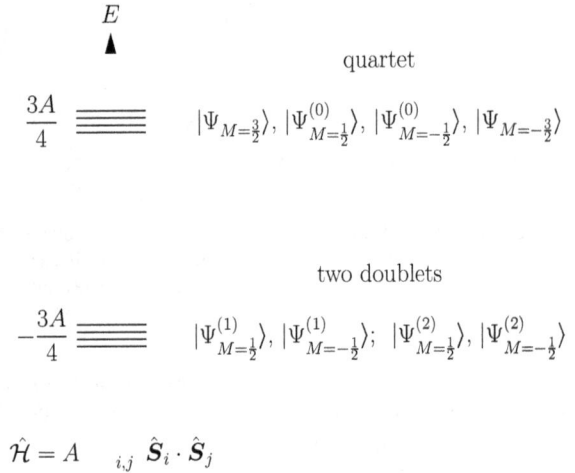

$$E$$

quartet

$$\frac{3A}{4} \;\;\equiv\equiv\equiv\;\; |\Psi_{M=\frac{3}{2}}\rangle,\; |\Psi^{(0)}_{M=\frac{1}{2}}\rangle,\; |\Psi^{(0)}_{M=-\frac{1}{2}}\rangle,\; |\Psi_{M=-\frac{3}{2}}\rangle$$

two doublets

$$-\frac{3A}{4} \;\;\equiv\equiv\equiv\;\; |\Psi^{(1)}_{M=\frac{1}{2}}\rangle,\; |\Psi^{(1)}_{M=-\frac{1}{2}}\rangle;\;\; |\Psi^{(2)}_{M=\frac{1}{2}}\rangle,\; |\Psi^{(2)}_{M=-\frac{1}{2}}\rangle$$

$$\hat{\mathcal{H}} = A \sum_{i,j} \hat{\boldsymbol{S}}_i \cdot \hat{\boldsymbol{S}}_j$$

It is interesting to consider the 8 states that make up these energy levels [18]. We can write these as follows

$$|\Psi_{M=\frac{3}{2}}\rangle = |\uparrow\uparrow\uparrow\rangle$$

$$|\Psi^{(k)}_{M=\frac{1}{2}}\rangle = \frac{1}{\sqrt{3}} \sum_{j=0}^{2} e^{2\pi i j k/3} C_3^j |\downarrow\uparrow\uparrow\rangle$$

$$|\Psi^{(k)}_{M=-\frac{1}{2}}\rangle = \frac{1}{\sqrt{3}} \sum_{j=0}^{2} e^{2\pi i j k/3} C_3^j |\uparrow\downarrow\downarrow\rangle$$

$$|\Psi_{M=-\frac{3}{2}}\rangle = |\downarrow\downarrow\downarrow\rangle \tag{2.44}$$

Two states are obvious. These are the 'ferromagnetic' configurations $|\uparrow\uparrow\uparrow\rangle$ and $|\downarrow\downarrow\downarrow\rangle$. The other six contributions have a single spin-flip with respect to these 'ferromagnetic configurations' and so are made up of states like $|\downarrow\uparrow\uparrow\rangle$. However, a state like $|\downarrow\uparrow\uparrow\rangle$ is not exchange symmetric or antisymmetric, so you have to make linear combinations such as $|\downarrow\uparrow\uparrow\rangle + |\uparrow\downarrow\uparrow\rangle + |\uparrow\uparrow\downarrow\rangle$. This is exactly what is achieved by the sums in (2.44) (and note that the $1/\sqrt{3}$ factor is simply a normalization). The operators C_3^j are threefold rotations of order j, and $j, k = 0, 1, 2$.

What is more, we can recover the chiral nature noted in the classical solutions. If we define a chirality operator $\hat{C}_z$ given by

$$\hat{C}_z = \frac{1}{4\sqrt{3}} \boldsymbol{S}_1 \cdot (\boldsymbol{S}_2 \times \boldsymbol{S}_3) , \tag{2.45}$$

then our states are eigenstates of the chirality operator

$$\hat{C}_z |\Psi_{M=\frac{3}{2}}\rangle = 0$$
$$\hat{C}_z |\Psi^{(0)}_{M=\frac{1}{2}}\rangle = 0$$
$$\hat{C}_z |\Psi^{(1)}_{M=\frac{1}{2}}\rangle = |\Psi^{(1)}_{M=\frac{1}{2}}\rangle$$
$$\hat{C}_z |\Psi^{(2)}_{M=\frac{1}{2}}\rangle = -|\Psi^{(2)}_{M=\frac{1}{2}}\rangle . \tag{2.46}$$

Note that the quantum numbers $k = 0, 1, 2$, and chirality $C_z = 0, \pm1$ describe the same states [18]. The states with non-zero chirality are in the pair of doublet ground states (for $A > 0$) while the excited state quartet has zero chirality.

2.4 Orbitals

2.4.1 Transition Metal Ions

The Heisenberg model only depends on the relative orientation of spins, not on their absolute orientation. It, therefore, seems to take no account of the lattice in which spins are embedded. However, spins are 'aware' of the lattice via the spin–orbit interaction. We now turn to consider the electronic orbitals that may be occupied in real systems. We will here particularly focus on transition metal compounds in which localized moments occur. (For metallic systems a band-like description would be more appropriate.)

Commonly occurring first-row transition metal ions are shown in Fig. 2.9 (taken from [19]), illustrating the range in occupancy of the $3d$ shell that can be obtained using transition metal ions in different oxidation states. I also show their electronic configuration for the case in which the ion experiences an octahedral crystal field (which splits the ten d-levels into a sixfold t_{2g} level and a fourfold e_g level [2]). Octahedral environments are very common in a wide variety of compound. This figure also shows that some ions are very particular about their oxidation state; for energetic reasons, Sc^{III} and Zn^{II} are the only stable states. On the other hand, an ion such as Mn can have a very large range of oxidation states from Mn^{II} (in MnO) to Mn^{VII} (in potassium permanganate $KMnO_4$). In fact, with clever chemistry even Mn^I is possible [20]! This range of occupancy leads to different spin and orbital moments of the ion and hence can be used to control the magnetic properties of the resulting system. For ions with d^4, d^5, d^6 and d^7 configuration, there is the possibility of both low-spin and high-spin states.

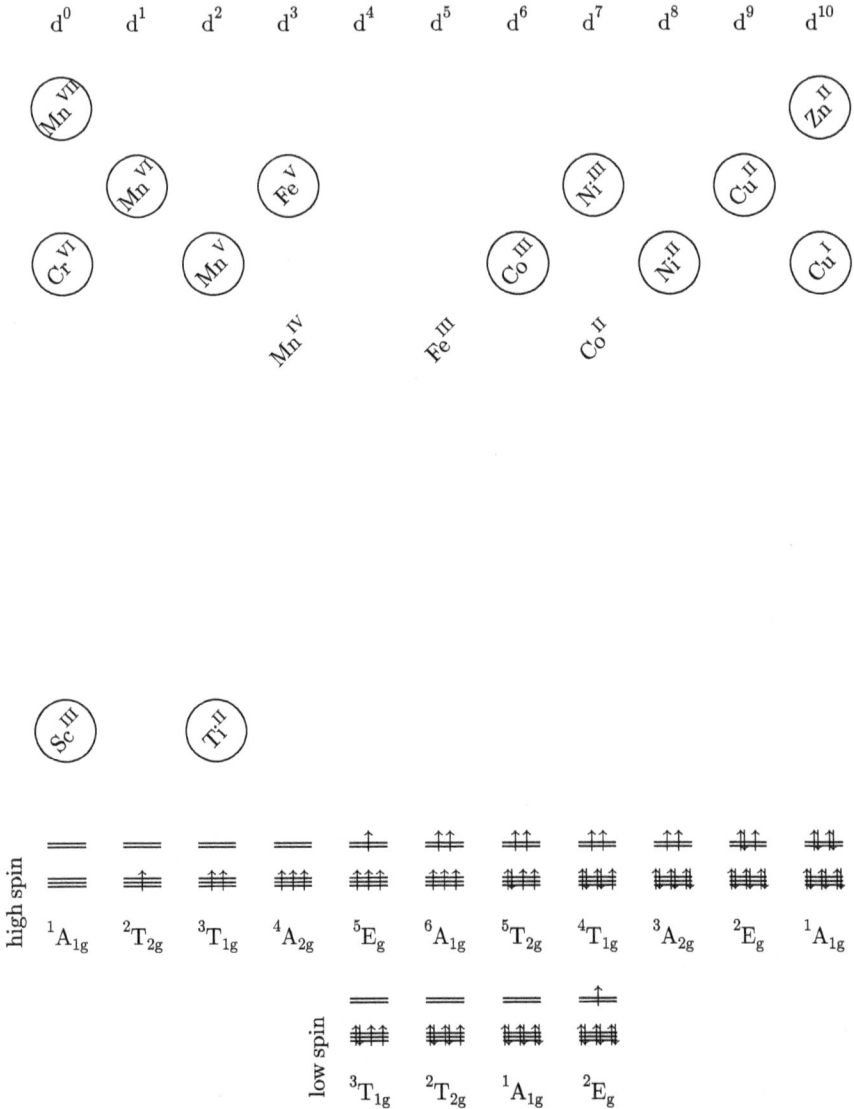

Fig. 2.9 Electronic states of the d-electrons in octahedrally coordinated first-row transition metal ions in commonly found oxidation states. The d-shell contains 10 electrons and so configurations $d^0, d^1, \ldots, d^{10}$ are possible, indicated by the vertical lines. Depending on the oxidation state chosen (and not all possibilities are chemically available) these configurations can be achieved using different first-row transition metal ions (Sc, Ti, V, Cr, Mn, Fe, Co, Ni, Cu, Zn). The oxidation state is indicated by the roman superscript, so that Cu^{II} is copper in its doubly ionized oxidation state (i.e. Cu^{2+}). An octahedral crystal field splits the ten d-levels into a sixfold t_{2g} level and a fourfold e_g level and the electron configurations are shown schematically at the bottom of the figure together with their description using the conventional spectroscopic notation ('A' signifies an orbitally non-degenerate state, 'E' doubly degenerate and 'T' triply degenerate). For ions with d^4, d^5, d^6 and d^7 configuration, there is the possibility of low-spin and high-spin configurations (Figure from [19])

2.4.2 Spin–Orbit Interaction and Crystal Fields

Now that we have the possibility of some partially filled d-levels, there can be magnetism. Next let us turn to the spin–orbit interaction which has the form

$$\lambda \hat{S} \cdot \hat{L} , \qquad (2.47)$$

so that the $\hat{S}$ operator acts on the spin part of the wave function, while the $\hat{L}$ operator acts on the spatial part of the wave function. If the states are approximately atomic states, and the spin–orbit interaction acts as a perturbation, one can focus on the $\lambda S^z L^z$ part and note that

$$\hat{L}^z = -i\hbar \frac{\partial}{\partial \phi} , \qquad (2.48)$$

which has eigenfunctions $e^{im\phi}$, i.e.

$$\hat{L}^z e^{im\phi} = -i\hbar \frac{\partial}{\partial \phi} e^{im\phi} = m\hbar e^{im\phi} . \qquad (2.49)$$

Now the crystal field is a real potential which is due to electrostatic fields from neighbouring ions. The eigenfunctions of the crystal field cannot be proportional to $e^{im\phi}$ because we require real solutions. (Recall the problem of particle in a box where the solutions are real and take the form $\cos kx$ or $\sin kx$, but not e^{ikx}, but of course we can make real functions by making linear combinations such as $e^{ikx} \pm e^{-ikx}$ and making a wave function out of something proportional to that.) To make a crystal field state we must, therefore, look for linear combinations of eigenfunctions such as

$$|\psi\rangle = e^{im\phi} \pm e^{-im\phi} . \qquad (2.50)$$

This kind of state though will automatically have zero angular momentum along the z-direction because it is made up of an equal contribution of a state with $\langle L^z \rangle = m\hbar$ and $\langle L^z \rangle = -m\hbar$. In fact, this idea works for all directions and

$$\langle \hat{L} \rangle = 0 . \qquad (2.51)$$

For example, if $l = 1$, there are three states with $m = 1, 0, -1$ with wave functions given by the spherical harmonics $Y_{1m}(\theta, \phi)$, and are therefore proportional to $\sin\theta\, e^{i\phi}$, $\cos\theta$ and $\sin\theta\, e^{-i\phi}$, respectively. We could write these states as $|1\rangle$, $|0\rangle$ and $|-1\rangle$. However, these are *not* the famous p-orbitals familiar from chemistry books. These arise in the formation of chemical bonds due to (real) electrostatic effects and must, therefore, be linear combinations of exactly the kind we are talking about. Thus the p-orbitals that line up along the x-, y- and z-directions [see Fig. 2.10a] are the zero-angular-momentum linear combinations given below:

(a)

(b)

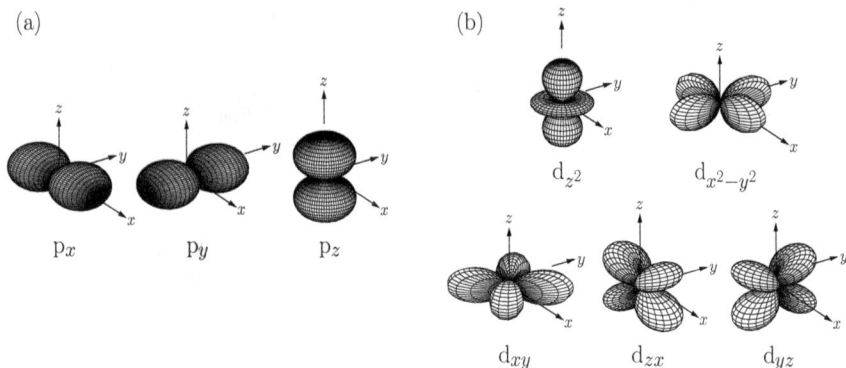

Fig. 2.10 Angular representation of **a** p-orbitals and **b** d-orbitals

$$|p_x\rangle = \frac{|1\rangle + |-1\rangle}{\sqrt{2}}$$

$$|p_y\rangle = \frac{|1\rangle - |-1\rangle}{\sqrt{2}i}$$

$$|p_z\rangle = |0\rangle . \tag{2.52}$$

What we see here is the phenomenon of quenching of angular momentum and it is interesting that it shows up even in the unfamiliar setting of the 'balloon animals' of p-orbitals. Of course in magnetism we are usually more interested in considering d-orbitals or f-orbitals but the same principles hold. For example, the equivalent equations for the d-orbitals [see Fig. 2.10b] are

$$|d_{xy}\rangle = \frac{|2\rangle - |-2\rangle}{\sqrt{2}i}$$

$$|d_{x^2-y^2}\rangle = \frac{|2\rangle + |-2\rangle}{\sqrt{2}}$$

$$|d_{yz}\rangle = \frac{-|1\rangle - |-1\rangle}{\sqrt{2}i}$$

$$|d_{zx}\rangle = \frac{-|1\rangle + |-1\rangle}{\sqrt{2}}$$

$$|d_{z^2}\rangle = |0\rangle . \tag{2.53}$$

2.4.3 Jahn–Teller Effect

It can sometimes be energetically favourable for, say, an octahedron to spontaneously distort as shown in Fig. 2.11 because the energy cost of increased elastic energy is balanced by a resultant electronic energy saving due to the distortion. This phe-

Fig. 2.11 The Jahn–Teller
effect for Mn^{3+} ($3d^4$). An
octahedral complex (left) can
distort (right), thus splitting
the t_{2g} and e_g levels. The
distortion lowers the energy
because the singly occupied
e_g level is lowered in energy.
The saving in energy from
the lowering of the d_{xz} and
d_{yz} levels is exactly balanced
by the raising of the d_{xy} level

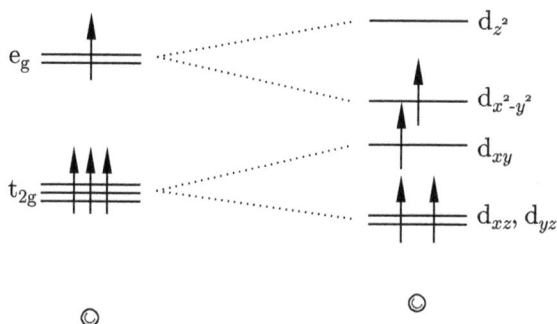

nomenon is known as the Jahn–Teller effect [21]. The distortion lowers the overall
energy by breaking an orbital degeneracy. For example, Mn^{3+} ions (which have a
configuration $3d^4$) in an octahedral environment show this kind of behaviour (see
Fig. 2.11) because the distortion can break the orbital degeneracy in the e_g levels. In
contrast, Mn^{4+} ions ($3d^3$) would not show this effect because there is no net lowering
of the electronic energy by a distortion.

To describe the effect, at least at the phenomenological level, we will assume that
the distortion of the system can be quantified by a parameter Q, which denotes the
distance of distortion along an appropriate normal mode coordinate. This gives rise
to an energy cost which is quadratic in Q and can be written as

$$E(Q) = \frac{1}{2}M\omega^2 Q^2 , \qquad (2.54)$$

where M and ω are, respectively, the mass of the anion and the angular frequency
corresponding to the particular normal mode. Clearly the minimum distortion energy
is zero and is obtained when $Q = 0$ (no distortion).

The distortion also raises the energy of certain orbitals while lowering the energy
of others. If all orbitals are either completely full or completely empty, this does not
matter since the overall energy is simply given by (2.54). However, in the cases of
partially filled orbitals this effect can be highly significant since the system can have a
net reduction in total energy. The electronic energy dependence on Q could be rather
complicated, but one can write it as a Taylor series in Q and provided the distortion
is small it is legitimate to keep only the term linear in Q. Let us, therefore, suppose
that the energy of a given orbital has a term either $+AQ$ or $-AQ$ corresponding to a
raising or a lowering of the electronic energy, where A is a suitable constant, assumed

to be positive. Then the total energy $E(Q)$ is given by the sum of the electronic energy and the elastic energy

$$E(Q) = \pm AQ + \frac{1}{2}M\omega^2 Q^2 , \qquad (2.55)$$

where the two possible choices of the sign of the AQ term give rise to two separate curves. If we consider only one of them we can find the minimum energy for that orbital using $\partial E/\partial Q = 0$ which yields a value of Q given by

$$Q_0 = \frac{A}{M\omega^2} \qquad (2.56)$$

and a minimum energy which is given by $E_{min} = -A^2/2M\omega^2$ which is less than zero. If only that orbital is full, then the system can make a net energy saving by spontaneously distorting.

LaMnO$_3$ contains Mn in the Mn^{3+} state which is a Jahn–Teller ion. LaMnO$_3$ shows A-type antiferromagnetic ordering. If a fraction x of the trivalent La^{3+} ions are replaced by divalent Sr^{2+}, Ca^{2+} or Ba^{2+} ions, holes are introduced on the Mn sites. This results in a fraction $1 - x$ of the Mn ions remaining as Mn^{3+} ($3d^4$, $t_{2g}^3 e_g^1$) and a fraction x becoming Mn^{4+} ($3d^4$, $t_{2g}^3 e_g^0$). When $x = 0.2$ the Jahn–Teller distortion vanishes and the system becomes ferromagnetic with a Curie temperature (T_C) around room temperature. Above T_C, the material is insulating and non magnetic, but below T_C, it is metallic and ferromagnetic. Particularly near T_C, the material shows an extremely large magnetoresistive effect which has been called colossal magnetoresistance.

The situation is actually more complicated because the carriers interact with phonons because of the Jahn–Teller effect. The strong electron–phonon coupling in these systems implies that the carriers are actually polarons above T_C, i.e. electrons accompanied by a large lattice distortion. These polarons are magnetic and are self-trapped in the lattice. The transition to the magnetic state can be regarded as an unbinding of the trapped polarons. There are other signatures of the electron–phonon couplings, including magnetic-field dependent structural transitions and charge ordering.

2.5 Conclusion

This chapter has discussed a number of important concepts in magnetism. Clearly this has just scratched the surface and for more details the reader should look elsewhere [2–5]. Nevertheless, these principles are helpful in understanding the wide variety of magnetic materials that are being studied, from frustrated magnets [22] to molecular magnets [19, 23] and from permanent magnetic materials [24] to spintronics [25].

References

1. J.H. van Leeuwen, Problèmes de la théorie électronique du magnétisme. J. Phys. Radium **2**, 361 (1921). https://doi.org/10.1051/jphysrad:01921002012036100
2. S. Blundell, *Magnetism in Condensed Matter* (Oxford University Press, New York, 2001). https://doi.org/10.1002/9781119280453
3. J.M.D. Coey, *Magnetism and Magnetic Materials* (Cambridge University Press, Cambridge, 2010). https://doi.org/10.1017/CBO9780511845000
4. N. Spaldin, *Magnetic Materials: Fundamentals and Applications* (Cambridge University Press, Cambridge, 2010). https://doi.org/10.1017/CBO9780511781599
5. D.C. Mattis, *The Theory of Magnetism Made Simple* (World Scientific. Singapore (2006). https://doi.org/10.1142/5372
6. M.A. Ruderman, C. Kittel, Indirect exchange coupling of nuclear magnetic moments by conduction electrons. Phys. Rev. **96**, 99 (1954). https://doi.org/10.1103/PhysRev.96.99
7. T. Kasuya, A theory of metallic ferro- and antiferromagnetism on Zener's model. Progr. Theoret. Phys. **16**, 45 (1956). https://doi.org/10.1143/PTP.16.45
8. K. Yosida, Magnetic properties of Cu-Mn alloys. Phys. Rev. **106**, 893 (1957). https://doi.org/10.1103/PhysRev.106.893
9. T. Lancaster, S.J. Blundell, *Quantum Field Theory for the Gifted Amateur* (Oxford Unversity Press. Oxford (2014). https://doi.org/10.1093/acprof:oso/9780199699322.001.0001
10. P.W. Anderson, *Theory of magnetic exchange interactions: exchange in insulators and semiconductors*, in *Solid State Physics*, ed. by F. Seitz, D. Turnbull, vol. 14 (Academic, New York, 1963), p. 99. https://doi.org/10.1016/S0081-1947(08)60260-X
11. P.W. Anderson, Antiferromagnetism. Theory of superexchange interaction. Phys. Rev. **79**, 350 (1950). https://doi.org/10.1103/PhysRev.79.350
12. J.B. Goodenough, Theory of the role of covalence in the perovskite-type manganites [La, $M(II)$]MnO$_3$. Phys. Rev. **100**, 564 (1955). https://doi.org/10.1103/PhysRev.100.564
13. J.B. Goodenough, An interpretation of the magnetic properties of the perovskite-type mixed crystals La$_{1-x}$Sr$_x$CoO$_{3-\lambda}$. J. Phys. Chem. Solids **6**, 287 (1958). https://doi.org/10.1016/0022-3697(58)90107-0
14. J. Kanamori, Superexchange interaction and symmetry properties of electron orbitals. J. Phys. Chem. Solids **10**, 87 (1959). https://doi.org/10.1016/0022-3697(59)90061-7
15. N.D. Mermin, H. Wagner, Absence of ferromagnetism or antiferromagnetism in one- or two-dimensional isotropic Heisenberg models. Phys. Rev. Lett. **17**, 1133 (1966). [Erratum: Phys. Rev. Lett. **17**, 1307 (1966).] https://doi.org/10.1103/PhysRevLett.17.1133
16. P.C. Hohenberg, Existence of long-range order in one and two dimensions. Phys. Rev. **158**, 383 (1967). https://doi.org/10.1103/PhysRev.158.383
17. S. Coleman, There are no Goldstone bosons in two dimensions. Math. Phys. **31**, 259 (1973). https://doi.org/10.1007/BF01646487
18. M. Trif, F. Trioiani, D. Stepanenko, D. Loss, Spin electric effects in molecular antiferromagnets. Phys. Rev. B **82**, 045429 (2000). https://doi.org/10.1103/PhysRevB.82.045429
19. S.J. Blundell, Molecular magnets. Contemp. Phys. **48**, 275 (2007). https://doi.org/10.1080/00107510801967415
20. E. Dixon, J. Hadermann, S. Ramos, A.L. Goodwin, M.A. Hayward, Mn(I) in an extended oxide: the synthesis and characterization of La$_{1-x}$Ca$_x$MnO$_{2+\delta}$ ($0.6 \leq x \leq 1$). J. Amer. Chem. Soc. **133**, 18397 (2011). https://doi.org/10.1021/ja207616c
21. H.A. Jahn, E. Teller, Stability of polyatomic molecules in degenerate electronic states - I. orbital degeneracy. Proc. R. Soc. London Ser. A **161**, 220 (1937). https://doi.org/10.1098/rspa.1937.0142
22. C. Lacroix, P. Mendels, F. Mila (ed.), *Introduction to Frustrated Magnetism (Materials, Experiments, Theory)*, Springer Series in Solid-State Sciences, vol. 164 (Springer, Berlin, 2011). https://doi.org/10.1007/978-3-642-10589-0
23. S.J. Blundell, F.L. Pratt, Organic and molecular magnets. J. Phys.: Condens. Matter **16**, R771 (2004). https://doi.org/10.1088/0953-8984/16/24/R03

3

XAS and PES: Basic Theory and Popular Computation Methods

Peter Krüger

Abstract The principles of X-ray absorption and photoemission spectroscopy calculations are introduced and the basics of electronic structure theory, including the Hartree–Fock approximation, density functional theory, its time-dependent version and quasiparticle theory are reviewed on an elementary level. Emphasis is put on polarization effects and the role played by electron correlation.

3.1 Introduction

In this chapter, the basic theory of X-ray absorption spectroscopy (XAS) and photoemission spectroscopy (PES) is introduced and popular computational methods are reviewed. Since XAS and PES mainly probe electronic excitations, a thorough understanding of electronic structure theory is mandatory. We shall review the standard theoretical methods for ground state electronic structure calculations, namely, Hartree–Fock (HF) and density functional theory (DFT). Among the various excited state theories, we focus on time-dependent DFT and briefly touch upon Green's function quasiparticle methods and the Bethe–Salpeter equation approach. We do not discuss ligand-field atomic multiplet theory, because this important method for transition metal L-edge calculations is covered in Chap. 4.

3.2 Light–Matter Interaction

As light is an electromagnetic wave, it interacts with all charged particles. In the visible to X-ray regime, the interaction with the electrons hugely dominates the interaction with the atomic nuclei. We shall therefore disregard the nuclear degrees

P. Krüger (✉)
Graduate School of Engineering and Molecular Chirality Research Center, Chiba University, Chiba 263-8522, Japan
e-mail: pkruger@chiba-u.jp

of freedom in the following, focusing on the electronic state of the system. When an X-ray photon impinges on an atom it can be either absorbed or scattered, energy, momentum and spin being, of course, conserved in the whole process. In absorption, the photon vanishes: all its energy is transferred to an electron which is excited to an empty state above the vacuum level. In scattering, the energy of the photon can remain the same [elastic (Thomson) scattering]; it can also be partly transferred to the atom (inelastic scattering) as in Compton scattering, which leads to the ejection of an electron, or in a Raman-like scattering, in which the energy lost by the photon brings the atoms in an excited state, without any ionization. Neglecting relativistic effects and treating the X-ray field classically, the light–electron interaction is obtained by replacing, in the electronic Hamiltonian, the electron momentum operator $\mathbf{p}$ by $\mathbf{p} - e\mathbf{A}/c$, where $\mathbf{A}(\mathbf{r}, t)$ is the vector potential of the light [1]. In Coulomb gauge ($\nabla \cdot \mathbf{A} = 0$), the interaction Hamiltonian then becomes

$$H_{\text{int}} = -\frac{e}{mc}\mathbf{A} \cdot \mathbf{p} + \frac{e^2}{2mc^2}A^2 . \tag{3.1}$$

The first-order term in A describes light absorption and stimulated emission while the second-order term is responsible for (non-resonant) light scattering. Here we focus on the absorption process and neglect the generally much weaker A^2 term. First-order perturbation theory (Fermi golden rule) leads to the following expression for the absorption intensity from an initial state $|i\rangle$ of energy E_i:

$$I(\omega) \propto \sum_f |\langle f|\mathbf{A} \cdot \mathbf{p}|i\rangle|^2 \delta(\hbar\omega - E_f + E_i) , \tag{3.2}$$

where the sum runs over all possible final states $|f\rangle$ with energy E_f. It is common to make the dipole approximation, i.e. to neglect the spatial variation of the X-ray field $\mathbf{A}(\mathbf{r})$. We may also replace the transition operator $\mathbf{A} \cdot \mathbf{p}$, by $\mathbf{e} \cdot \mathbf{r}$, where $\mathbf{e} = \mathbf{A}/|\mathbf{A}|$ is the light polarization vector and the change from $\mathbf{p}$ to $\mathbf{r}$ is possible by exploiting commutation relations between $\mathbf{r}, \mathbf{p}$ and H, and the fact that $|i\rangle$ and $|f\rangle$ are energy eigenstates [2]. Equation (3.2) is often interpreted in a single-particle picture, in which case $|i\rangle$ is an atomic core state and $|f\rangle$ are unoccupied states above the Fermi level. However, electrons interact with each other through the Coulomb interaction, such that the excitation of one electron affects the motion of the others. Therefore, the correct use of (3.2) is in a many-particle sense, where $|i\rangle = |\Phi_g\rangle$ is the many-electron ground state, and $|f\rangle = |\Phi_f\rangle$ are many-electron excited states with a core hole. Putting the constants we have

$$I(\omega) = 4\pi^2\alpha\hbar\omega \sum_f |\langle\Phi_f|\mathbf{e} \cdot \sum_j \mathbf{r}_j|\Phi_g\rangle|^2 \delta(\hbar\omega - E_f + E_g) , \tag{3.3}$$

where $\alpha = e^2/\hbar c$ is the fine structure constant and j counts the electrons. Having established the expressions of the absorption intensity, the remaining task is to calculate the eigenstates of the (unperturbed) electronic system, $|\Phi_g\rangle$ and $|\Phi_f\rangle$, and their energies. Thus, the main theoretical problem of XAS is the accurate description of the electronic structure of the system, both for the ground and core-excited states. We, therefore, start by reviewing the basics of (ground state) electronic structure theory before turning to the specific methods for handling core-excited states.

3.3 Ground State Electronic Structure Theory

Consider N electrons interacting with each other and the atomic nuclei. Following the Born–Oppenheimer approximation, we neglect the coupling between the nuclear and electronic dynamics. For the electronic problem, this means that the nuclei are at fixed positions and can be described by a static external potential $V_{\text{ext}}(\mathbf{r})$. The electronic Hamiltonian is then given by

$$H = T + V_{\text{ext}} + V_{ee} = \sum_i \frac{1}{2} \nabla_i^2 + \sum_i V_{\text{ext}}(\mathbf{r}_i) + \sum_{i<j} \frac{1}{|\mathbf{r}_i - \mathbf{r}_j|} , \qquad (3.4)$$

where i, j count the electrons and atomic units are used ($\hbar = m = e = 1$). The kinetic energy T and the external potential V_{ext} are one-particle operators whereas the electron–electron interaction V_{ee} is a two-particle operator. Because of V_{ee}, the electronic motion is correlated and the many-electron problem cannot be solved exactly (except for a few electrons). Drastic approximations need to be made. The most important ground state electronic structure methods are HF and DFT.

3.3.1 Hartree–Fock Approximation

Historically, the first accurate electronic structure method is the Hartree–Fock approximation (HFA) [3]. It is still widely used for single molecule calculations and as a starting point for more advanced schemes. The basic assumption of the HFA is that the many-electron ground state wave function is a Slater determinant, i.e. an antisymmetrized product of single-electron states (spin-orbitals). By applying the Rayleigh–Ritz variational principle, the HF equations are obtained, whose solutions are the HF orbitals $\phi_n(\mathbf{r})$ and energies ϵ_n. For convenience, we suppress the spin part of the single-particle wave functions. The HF equations are

$$\left(-\frac{1}{2} \nabla^2 + V_{\text{ext}}(\mathbf{r}) + V_{\text{H}}(\mathbf{r}) + V_X \right) \phi_n(\mathbf{r}) = \epsilon_n \phi_n(\mathbf{r}) . \qquad (3.5)$$

This is a one-electron Schrödinger equation where the pair-wise electron–electron interaction is replaced by an effective potential $V_{\text{H}} + V_X$.

$$V_H(\mathbf{r}) = \sum_m^{occ} \int d\mathbf{r}' \frac{|\phi_m(\mathbf{r}')|^2}{|\mathbf{r} - \mathbf{r}'|} = \int d\mathbf{r}' \frac{n(\mathbf{r}')}{|\mathbf{r} - \mathbf{r}'|} \tag{3.6}$$

is called the Hartree potential and corresponds to the classical electrostatic potential due to the electronic charge density $n(\mathbf{r}) = \sum_m^{occ} |\phi_m(\mathbf{r})|^2$ of the occupied orbitals.

$$V_X \phi_n(\mathbf{r}) = - \sum_m^{same\ spin} \int d\mathbf{r}' \frac{\phi_m^*(\mathbf{r}')\phi_m(\mathbf{r})\phi_n(\mathbf{r}')}{|\mathbf{r} - \mathbf{r}'|} \tag{3.7}$$

is the exchange potential which is due to the electron–electron interaction together with the antisymmetry of Slater determinants under permutation of two electrons. V_X is a non-local potential and has no classical analogue. Both V_H and V_X are static 'mean-field' potentials, obtained from the time-averaged orbital motion of the electrons. Dynamical effects are neglected. The exchange interaction induces some correlation between electrons of same spin, which avoid each other due to the Pauli principle. Correlation between electrons of opposite spin is completely absent in the HFA. By definition, the difference between the exact ground state and the HF ground state is called the electron correlation effect (even though mathematically speaking, there is correlation between same spin electrons in the HFA).

There are various methods to take account of electron correlation, often termed collectively as 'post-HF' methods in the chemical literature. The conceptually most simple way to include electron correlation is the configuration interaction (CI) method. In CI, a set of Slater determinants is generated from the HF ground state by (multiple) particle–hole excitations. The CI wave function is a linear combination of these many-electron basis states and the coefficients and total energy levels are determined variationally by diagonalizing the Hamiltonian in this sub-space. CI can be very accurate for atoms and small molecules, but cannot directly be applied to large molecules and materials because the number of Slater determinants grows exponentially with system size. For X-ray absorption spectra, CI effects, i.e. mixing between Slater determinants, are especially strong at transition metal L-edges and lanthanide M-edges, which correspond to excitations into the localized $3d$- and $4f$-orbitals. For these spectra, CI must be taken into account. This can be done with the ligand-field multiplet method which is based on CI of a single atom or a very small cluster (see Chap. 4 for details).

3.3.2 Density Functional Theory

Nowadays, most electronic structure methods are based on DFT [4, 5]. In DFT, one does not try to find approximations to the many-electron wave function. Instead, the idea is to directly find the exact electronic density $n(\mathbf{r})$ and total energy, which is expressed as a functional of the density. DFT is based on two theorems due to Hohenberg and Kohn [4] about the (non-degenerate) ground state of the interacting, inhomogeneous electron gas. The first theorem states that the external potential V_{ext} is uniquely determined by the ground state electronic density $n(\mathbf{r})$ and that the total energy E (minus the external potential energy) is a unique and universal functional of n.[1] The consequence of the theorem is that knowledge of the ground state density alone is, in principle, sufficient to determine all properties of the system. The second theorem states that the exact ground state density $n_0(\mathbf{r})$ minimizes the total energy functional $E[n]$ in the space of all possible functions $n(\mathbf{r})$. Thus, approximations to E and n can be found variationally.

The Hohenberg–Kohn theorems are exact mathematical theorems. If the universal functional $E[n]$ were known, DFT would yield the exact total energy and electron density of the interacting electron system. But the exact functional $E[n]$ is unknown. Various approximate functionals have been proposed such as the local density approximation (LDA), generalized gradient approximations (GGA) and hybrid functionals, i.e. mixtures of GGA and HF exchange. In practical DFT calculations, Kohn–Sham theory is employed, which introduces an auxiliary, non-interacting system which, by definition, has the same electronic density as the real, interacting system. In the auxiliary system, the external potential is called Kohn–Sham potential V_{KS} and it is the sum of the true external (nuclei) potential and an effective one-electron potential which replaces the electron–electron interaction. The Kohn–Sham potential V_{KS} is given by the functional derivative of the total energy functional $E[n]$ with respect to the electron density. The effective electron–electron potential is written as the sum of the Hartree potential V_H and a rest, which is called exchange–correlation potential V_{XC}. As the exact energy functional $E[n]$ is unknown, so is V_{XC}, and the actual expression depends on the approximation used (LDA, GGA, hybrid).

As the Kohn–Sham auxiliary system is non-interacting, its eigenstates are Slater determinants made of orbitals ϕ_n that are solutions of the Kohn–Sham equations

$$\left(-\frac{1}{2}\nabla^2 + V_{ext}(\mathbf{r}) + V_H(\mathbf{r}) + V_{XC}\right)\phi_n(\mathbf{r}) = \epsilon_n\phi_n(\mathbf{r}) . \tag{3.8}$$

The Kohn–Sham equations (3.8) are similar to the HF equations (3.5) except that the exchange potential V_X is replaced by the exchange–correlation potential V_{XC} and the expression of the total energy as a function of the orbitals is different.

DFT takes account of electron correlation through V_{XC} and generally performs better than the HFA for ground state properties. This is, however, not necessarily true for excited states for which DFT should, in principle, not be used, because the Kohn–Sham orbitals and levels ϵ_n describe the auxiliary system and have, strictly speaking, no direct physical meaning for the real system. In practice, however, the orbitals and energy levels are used in the same way as the HF orbitals, namely, as a first-order approximation for the one-electron or one-hole excitations of the system.

---

[1] The opposite is obvious because when V_{ext} is fixed, the Hamiltonian is known and so all properties, including the electronic density, are determined.

3.4 Absorption Spectra in the Independent Particle Approximation

Recalling (3.3), the absorption intensity is determined by the transition amplitude $M_{fg} = \langle \Phi_f | \mathbf{e} \cdot \sum_n \mathbf{r}_n | \Phi_g \rangle$. If both Φ_g and Φ_f are Slater determinants made of orbitals which are eigenstates of the same one-electron Hamiltonian, then it is easy to see that M_{fg} reduces to a one-particle transition matrix element between the core orbital $|\phi^c\rangle$ with energy ϵ_c and an unoccupied orbital $|\phi_k\rangle$ with energy ϵ_k and (3.3) simplifies to

$$I(\omega) = 4\pi^2 \alpha \hbar \omega \sum_k^{\epsilon_k > \epsilon_F} |\langle \phi_k | \mathbf{e} \cdot \mathbf{r} | \phi^c \rangle|^2 \delta(\hbar \omega - \epsilon_k + \epsilon_c) . \tag{3.9}$$

This is the basic equation of XAS in the independent particle approximation.

So far we have implicitly assumed that ϕ_k are the unoccupied orbitals of a ground state calculation. However, Φ_f is an excited state with a core hole. The core hole acts as a local positive charge which modifies the effective potential ($V_H + V_{X/XC}$) and so the best Slater determinant for Φ_f is made of a different set of orbitals $\tilde{\phi}_k$ than the ground state orbitals ϕ_k. Accordingly, better results are usually obtained with 'relaxed' orbitals $\tilde{\phi}_k$, corresponding to a constraint HF or DFT calculation with a core hole. As core holes are localized on one atomic site, the symmetry of the system is generally lowered in a core hole calculation and the computational cost increases. For crystals, in particular, a supercell calculation is needed in order to effectively separate the artificially repeated core hole sites. In the following, we shall write ϕ_k regardless for relaxed and unrelaxed orbitals.

3.4.1 Dipole Selection Rules and Density of States

For the calculation of the dipole transition matrix elements $\langle \phi_k | \mathbf{e} \cdot \mathbf{r} | \phi^c \rangle$, it is useful to expand the states $|\phi_k\rangle$ in a spherical harmonics basis centred at the atomic sites $\mathbf{R}_i$. By doing so, the dipole transition selection rules known from atomic physics can be exploited. This simplifies the calculation and yields an interpretation of the spectra in terms of projected density of states as we shall see. We write

$$\phi_k = \sum_{ilm} B^k_{ilm} \chi_{ilm} , \quad \chi_{ilm}(\mathbf{r}) = R_{il}(r_i) Y_{lm}(\mathbf{r}_i) , \tag{3.10}$$

where $\mathbf{r}_i \equiv \mathbf{r} - \mathbf{R}_i$, Y_{lm} are spherical harmonics, R_{il} radial functions and B_{ilm} complex coefficients. The core orbital is localized at some site (i_c). Therefore, only orbitals χ_{ilm} with $i = i_c$ give a non-zero contribution to the matrix element. Next we write the dipole operator as a spherical tensor product $\mathbf{e} \cdot \mathbf{r} = \sum_q (-1)^q e_{-q} r_q$, where $q = 0, \pm 1$ are the spherical components of a vector $\mathbf{a}$, given by $a_0 = a_z$,

$a_\pm = (\mp a_x - i a_y)/\sqrt{2}$. The angular integrals of the matrix elements can then be simplified with the help of the Wigner–Eckart theorem [2]

$$\langle n'l'm'|r_q|nlm\rangle = (-1)^{l'-m'} \begin{pmatrix} l' & 1 & l \\ -m' & q & m \end{pmatrix} \langle n'l'||r||nl\rangle , \qquad (3.11)$$

where $(\,\vdots\,\vdots\,)$ are Wigner-3j symbols and $\langle n'l'||r||nl\rangle$ are reduced matrix elements, which are independent of m, m', q. The Wigner-3j symbol is non-zero only for $l' = l \pm 1$ and $m' = m + q$. These are the dipole selection rules. For example, for K-edge spectra $l = m = 0$ and thus only $l' = 1$, i.e. p-type final states can be reached. If polarized light is used we further have $m' = q$, e.g. in z-polarization only p_z states are probed. We thus see that XAS is a local probe of the unoccupied electronic states, where different orbital symmetries can be projected out by appropriately choosing the absorption edge l value and the light polarization q.

Using the expansion (3.10) and the dipole selection rules (3.11), we find for the transition matrix elements from a core orbital ϕ^c_{ilm}

$$\langle \phi_k|r_q|\phi^c_{ilm}\rangle = \sum_\pm B^{k*}_{i,l\pm1,m+q}\langle \chi_{i,l\pm1,m+q}|r_q|\phi^c_{ilm}\rangle . \qquad (3.12)$$

The absorption intensity (3.9), dropping constants, from a core shell with angular momentum l, located at site i, for light polarization q becomes

$$I_q(\omega) = \sum_{km} \left| \sum_\pm B^{k*}_{i,l\pm1,m+q}\langle \chi_{i,l\pm1,m+q}|r_q|\phi^c_{ilm}\rangle \right|^2 \delta(\hbar\omega - \epsilon_k + \epsilon_c) . \qquad (3.13)$$

As defined in (3.10), the orbitals $R_{il}(r)$ and the expansion coefficients B are, in principle, energy dependent. This is the choice in multiple scattering theory which allows a minimal basis set (one orbital for each site and l). In the following, we neglect the slow energy dependence of the radial waves. We then obtain

$$I_q(\omega) \approx \sum_{m,a,b=\pm} M^{a*}_{ilm,q}M^b_{ilm,q} \sum_k B^k_{il_am+q}B^{k*}_{il_bm+q} \delta(\hbar\omega - \epsilon_k + \epsilon_c) , \qquad (3.14)$$

where $l_a, l_b = l \pm 1$ and $M^b_{ilm,q} = \langle \chi_{i,l_b,m+q}|r_q|\phi^c_{ilm}\rangle$. We introduce the local, orbital projected density of states matrix

$$\rho_{ilm,l'm'}(\epsilon) = \sum_k \langle \chi_{ilm}|\phi_k\rangle\delta(\epsilon - \epsilon_k)\langle \phi_k|\chi_{il'm'}\rangle = \sum_k B^k_{ilm}\delta(\epsilon - \epsilon_k)B^{k*}_{il'm'} . \qquad (3.15)$$

Note that the usual partial density of states (DOS) is given by the diagonal elements ($lm = l'm'$). So (3.14) can be written as

$$I_q(\omega) \approx \sum_{m,a,b=\pm} M^{a*}_{ilm,q} M^{b}_{ilm,q} \rho_{il_am+q,l_bm+q}(\hbar\omega + \epsilon_c) . \qquad (3.16)$$

We see that the absorption intensity is a weighted sum of a few partial DOS components with angular momentum $l_a = l \pm 1$. For high enough symmetry, the interference terms $a \neq b$ vanish, leaving only the diagonal, usual partial DOS. In some cases, e.g. for the linear dichroism at the sulphur $L_{2,3}$-edges in MoS_2, it was found that interference between $p \rightarrow s$ and $p \rightarrow d$ transitions is non-negligible [6]. For the special case of s-wave core states (K, L_1, M_1 edges) where $l = m = 0$, the selection rules (3.11) give $l' = 1, m = q$, such that the absorption spectrum for q-polarized light is directly proportional to the p_q-DOS (where $q = 0, \pm 1$ or $q = x, y, z$).

In this section, we have seen that in the independent particle and dipole approximation, the X-ray absorption spectra are approximately given by a weighted sum of partial DOS with momenta $l \pm 1, m + q$. The weighting factors are local transition matrix elements and reflect the light polarization and orbital symmetry. As a consequence, XAS can be used to probe the unoccupied DOS of the material in a site and orbital-resolved way, which gives detailed insight into the local bonding properties [7].

3.5 Absorption Spectra in Linear Response TDDFT

3.5.1 Time-Dependent Density Functional Theory

DFT is a ground state theory whose application to excited states is ill-founded. However, a large class of excitations can be computed using the time-dependent version of DFT. Time-dependent DFT (TDDFT) is the generalization of standard DFT to time-dependent external potentials $V_{ext}(\mathbf{r}, t)$. It was pioneered by Zangwill and Soven in 1980 [8], who developed a linear response theory for optical absorption spectroscopy of atoms using a time-dependent version of the LDA. In 1984, Runge and Gross [9] generalized the Hohenberg–Kohn theorems of DFT to the case of time-dependent systems, thus putting TDDFT on a rigorous theoretical ground. TDDFT has been applied to XAS of solids for the first time in 1998 by Schwitalla and Ebert [10] and to molecules in 2003 by Stener et al. [11].

The problem at hand is to find the time-dependent electron density $n(\mathbf{r}, t)$ of an interacting electron system subject to a time-dependent external field. In TDDFT, the exact time-dependent electron density $n(\mathbf{r}, t)$ can, in principle, be found from the knowledge of the external field, the universal energy functional $E[n(\mathbf{r}, t)]$ and the initial density $n(\mathbf{r}, 0)$. Linear response functions, including absorption coefficients, can be expressed as integrals over the electron density change induced by a time-dependent external field. Thus, if the exact functional $E[n(\mathbf{r}, t)]$ were known, TDDFT would allow to obtain exact absorption spectra. As in the case of time-independent

DFT, however, the exact functional is unknown. Moreover, finding good approximate functionals is even more difficult in TDDFT than in standard DFT.

3.5.2 Linear Response Theory

Here we shall outline the theory of absorption spectroscopy in linear response following Zangwill and Soven [8]. We consider an interacting electron system as described by the unperturbed Hamiltonian H in (3.4), and try to find its response to a time-dependent applied field $\varphi_{ext}(\mathbf{r}, t)$ such as the electromagnetic field of an X-ray beam. The perturbation Hamiltonian is written as

$$H'(t) = \int \varphi_{ext}(\mathbf{r}, t) n(\mathbf{r}, t) d\mathbf{r} , \qquad (3.17)$$

where $n(\mathbf{r}, t)$ is the electron density. It differs from the density of the unperturbed system $n^0(\mathbf{r})$ by the induced density

$$\delta n(\mathbf{r}, t) = n(\mathbf{r}, t) - n^0(\mathbf{r}) . \qquad (3.18)$$

The fundamental assumption of *linear* response theory is that the response of the system, δn, is proportional to the applied field φ_{ext}, i.e.

$$\delta n(\mathbf{r}, t) = \int d\mathbf{r}' dt' \chi(\mathbf{r}, \mathbf{r}', t - t') \varphi_{ext}(\mathbf{r}', t') , \qquad (3.19)$$

where χ, the response function, is an intrinsic property of the unperturbed system. In the frequency domain, this relation reads

$$\delta n(\mathbf{r}, \omega) = \int d\mathbf{r}' \chi(\mathbf{r}, \mathbf{r}', \omega) \varphi_{ext}(\mathbf{r}', \omega) . \qquad (3.20)$$

It can be shown that χ is given by the retarded density–density Green's function

$$\chi(\mathbf{r}, \mathbf{r}', t - t') = -i\theta(t - t')\langle 0|[\hat{n}(\mathbf{r}, t), \hat{n}(\mathbf{r}', t')]|0\rangle , \qquad (3.21)$$

where $\hat{n}(t) = e^{iHt}\hat{n}e^{-iHt}$ is the density operator in Heisenberg representation, $|0\rangle$ is the exact ground state of H with energy E_0, $[,]$ denotes the commutator and $\theta(x)$ is the Heaviside step function $[\theta(x) = 1$ for $x > 0$ and $\theta(x) = 0$ for $x < 0]$. By inserting a complete set of excited states $\sum_m |m\rangle\langle m|$ and performing a time-frequency Fourier transformation, we obtain the following exact ('Lehmann') representation:

$$\chi(\mathbf{r}, \mathbf{r}', \omega) = \sum_m \frac{\langle 0|\hat{n}(\mathbf{r})|m\rangle\langle m|\hat{n}(\mathbf{r}')|0\rangle}{\hbar\omega - E_m + E_0 + i\eta} - \sum_m \frac{\langle 0|\hat{n}(\mathbf{r}')|m\rangle\langle m|\hat{n}(\mathbf{r})|0\rangle}{\hbar\omega + E_m - E_0 + i\eta} , \qquad (3.22)$$

where $|m\rangle$ are excited states with energy E_m and η is an infinitesimal positive number. The exact eigenstates and energies of the interacting electron systems are unknown, so (3.22) cannot be evaluated directly. For a non-interacting electron gas, however, all eigenstates are Slater determinants, and (3.22) can be calculated. The only excitations which give non-zero matrix elements are single particle–hole excitations $|m\rangle = c_p^+ c_h |0\rangle$ with energy $\epsilon_p - \epsilon_h$, where p and h label states above and below the Fermi level, respectively. This gives the response function in the independent particle approximation

$$\chi_0(\mathbf{r}, \mathbf{r}', \omega) = \sum_{hp} \frac{\phi_h^*(\mathbf{r})\phi_p(\mathbf{r})\phi_p^*(\mathbf{r}')\phi_h(\mathbf{r}')}{\hbar\omega - \epsilon_p + \epsilon_h + i\eta} - [p \leftrightarrow h] . \tag{3.23}$$

If the electrons did not interact we would have $\chi = \chi_0$. But they do interact. In TDDFT, the interaction is handled as in DFT, by introducing an auxiliary, non-interacting system with the same electron density $n(\mathbf{r}, t)$ which corresponds to a time-dependent effective potential. In the real system, the density change $\delta n(\mathbf{r}, t)$ is induced by the external perturbation $\varphi_{\text{ext}}(\mathbf{r}, t)$. In the auxiliary system, however, the density $n(\mathbf{r}, t)$ corresponds to the sum of the Kohn–Sham potential V_{KS} and the perturbation φ_{ext}. As the Kohn–Sham potential depends on the density, a density change δn gives rise to an induced field $\varphi_{\text{ind}}(\mathbf{r}, t) = \delta V_{\text{KS}}[n(\mathbf{r}, t)]$. Thus, the density change $\delta n(\mathbf{r}, t)$ is due not only to the true external potential φ_{ext} but also to the induced field φ_{ind}. Note that there is a feedback effect: $\varphi_{\text{ext}} \to \delta n \to \varphi_{\text{ind}} \to \delta^2 n \to \delta\varphi_{\text{ind}} \ldots$, so we need to solve for δn and φ_{ind} self-consistently. Further, in linear response theory, a linear relation between the induced charge density and the induced field is assumed

$$\varphi_{\text{ind}}(\mathbf{r}t) = \int d\mathbf{r}' dt' K(\mathbf{r}t, \mathbf{r}'t')\delta n(\mathbf{r}'t') \tag{3.24}$$

which defines the interaction kernel K. We have $\varphi_{\text{ind}} \equiv \delta V_{\text{KS}} = \delta V_{\text{H}} + \delta V_{XC}$, where

$$\delta V_{\text{H}}(\mathbf{r}t) = \int d\mathbf{r}' \frac{\delta n(\mathbf{r}'t)}{|\mathbf{r} - \mathbf{r}'|} , \quad \delta V_{XC}(\mathbf{r}t) = \int d\mathbf{r}' dt' \frac{\delta V_{XC}(\mathbf{r}t)}{\delta n(\mathbf{r}'t')}\delta n(\mathbf{r}'t') . \tag{3.25}$$

The total time-dependent perturbation in the auxiliary system is often called the 'local field', $\varphi_{\text{loc}} = \varphi_{\text{ext}} + \varphi_{\text{ind}}$. As the electrons of the auxiliary system are independent, they respond to the perturbation φ_{loc} with the free response function χ_0, i.e. $\delta n(\mathbf{r}t) = \int d\mathbf{r}' dt' \chi_0(\mathbf{r}, \mathbf{r}', t - t')\phi_{\text{loc}}(\mathbf{r}'t')$. By construction, the charge densities of the real and auxiliary systems are the same, so we have

$$\chi \, \varphi_{\text{ext}} = \delta n = \chi_0 \, \varphi_{\text{loc}} = \chi_0(\varphi_{\text{ext}} + K\delta n) = \chi_0(1 + K\chi)\varphi_{\text{ext}} , \tag{3.26}$$

where arguments and integration symbols have been suppressed to simplify the notation. Since φ_{ext} is arbitrary, we have

$$\chi = \chi_0 + \chi_0 K \chi \quad \Leftrightarrow \quad \chi = (\chi_0^{-1} - K)^{-1} . \tag{3.27}$$

So the full response function χ can be calculated from free response function χ_0 and the kernel K, by iteration or inversion. Equivalently one can calculate the local potential directly by iteration of $\varphi_{\text{loc}} = \varphi_{\text{ext}} + K\chi_0\varphi_{\text{loc}}$ [8]. The problem is that the exchange–correlation part of the kernel

$$K_{XC}(\mathbf{r}t, \mathbf{r}'t') = \frac{\delta V_{XC}(\mathbf{r}t)}{\delta n(\mathbf{r}'t')} \tag{3.28}$$

is not known exactly. The adiabatic approximation consists in using a static exchange–correlation potential, which may be taken from standard time-independent DFT. In this case, $K_{XC} = [\delta V_{XC}(\mathbf{r})/\delta n(\mathbf{r}')]\delta(t - t')$ such that $\varphi_{\text{ind}}(t)$ changes instantaneously with $\delta n(t)$. As a result, $K(\mathbf{r}, \mathbf{r}', \omega)$ is frequency independent and dynamical screening is neglected. X-ray fields correspond to fast oscillations, so neglecting dynamical effects is questionable.

3.5.3 Absorption Spectra

The optical absorption coefficient is essentially the imaginary part of the response function χ as we shall show now. We consider an electromagnetic wave given by $\mathbf{E}(\mathbf{r}, t) = \mathbf{e}E_0 e^{i\mathbf{q}\cdot\mathbf{r}-i\omega t}$, where $\mathbf{e}$ is the light polarization vector, not to be confused with the electric charge e. The induced electrical polarization is $\mathbf{P}(\mathbf{r}, t) = -e\delta n(\mathbf{r}, t)\mathbf{r}$ and so the change in energy density is $-\mathbf{E} \cdot \mathbf{P} = e\mathbf{E} \cdot \mathbf{r}\delta n$. In the dipole approximation, $\mathbf{E}(\mathbf{r}, t) \approx \mathbf{e}E_0 e^{-i\omega t}$ and the perturbation in (3.17) is given by

$$\varphi_{\text{ext}}(\mathbf{r}, \omega) = eE_0\mathbf{e} \cdot \mathbf{r} . \tag{3.29}$$

The total induced dipole moment is

$$\boldsymbol{\mu}(\omega) = -e\int \mathbf{r}\delta n(\mathbf{r}, \omega)d\mathbf{r} = -e^2 E_0 \int \mathbf{r}\chi(\mathbf{r}, \mathbf{r}', \omega)\mathbf{e} \cdot \mathbf{r}'d\mathbf{r}d\mathbf{r}' , \tag{3.30}$$

and the absorbed energy is $\text{Re}\{\mathbf{E} \cdot d\boldsymbol{\mu}/dt\}$ or equivalently $\text{Im}\{\omega\mathbf{E} \cdot \boldsymbol{\mu}(\omega)\}$. The absorption coefficient $\sigma(\omega)$ is the absorbed energy divided by E_0^2, which yields

$$\sigma(\omega) = -4\pi\alpha\hbar\omega \int d\mathbf{r}d\mathbf{r}'\mathbf{e} \cdot \mathbf{r} \, \text{Im}\chi(\mathbf{r}, \mathbf{r}', \omega)\mathbf{e} \cdot \mathbf{r}' . \tag{3.31}$$

This expression of the absorption coefficient is fully equivalent to (3.3). In the independent particle approximation, we put $\chi \to \chi_0$ and obtain from (3.23)

$$\sigma_0(\omega) = 4\pi^2\alpha\hbar\omega \sum_{hp} |\langle\phi_p|\mathbf{e} \cdot \mathbf{r}|\phi_h\rangle|^2\delta(\hbar\omega + \epsilon_h - \epsilon_p) \tag{3.32}$$

in agreement with (3.9).

In summary, TDDFT with linear response provides a rigorous and efficient framework for calculating absorption spectra. Compared to the independent particle approximation, TDDFT takes the screening of the electromagnetic field into account by introducing the induced field $\varphi_{\text{ind}}(t)$ which is calculated self-consistently with the density change $\delta n(t)$. In practice, the problem is to find good approximations for the unknown exchange–correlation kernel K_{XC}. The Hartree part alone, i.e. putting $K_{XC} = 0$, yields the well-known random-phase approximation (RPA) [10]. Apart from the single particle–hole excitations included in χ_0, the RPA can describe plasmon excitations, i.e. collective oscillations of electron gas, which can be observed, for example, as satellite peaks in core-level photoemission spectra. The RPA kernel also gives rise to a redistribution of spectral weight between different transitions. This may strongly change the peak intensity ratio, e.g. between the L_2 and L_3 white lines in transition elements [10]. In adiabatic TDDFT, K_{XC} can be obtained from standard DFT [8], but such static approximations to K_{XC} do not improve much over the RPA [10, 12]. It appears that complex configuration mixing such as multiplet excitations cannot be described by the common, adiabatic kernels. Going beyond the adiabatic approximation is difficult, but some non-adiabatic kernels have been proposed and applied to the X-ray absorption problem [13].

3.6 Photoemission Spectroscopy

PES is probably the most direct way of probing the electronic structure of materials. In a PES, light is shone on a surface and the kinetic energy, and possibly exit angle and spin, of the emitted electrons is measured. In core-level PES, electrons from the inner atomic shells are excited. As these levels are element specific, core-level photoemission is a powerful tool for chemical analysis.

Angle-resolved core-level photoemission from crystal surfaces is known as X-ray photoelectron diffraction [14]. The photoelectron wave spreads from the core hole site and is diffracted by the neighbouring atoms. Analysis of the diffraction pattern gives precise information about the local structure around the atoms of a given chemical species. X-ray photoelectron diffraction can be well modelled with real-space single or multiple scattering theory on a finite cluster of atoms.

3.6.1 Angle-Resolved Photoemission Spectroscopy

Angle-resolved photoemission spectroscopy (ARPES) is the major method for measuring energy band dispersion (the 'band structure') of crystals. An intuitive picture of ARPES is provided by the three-step model [15]. The three steps are as follows:

1. Photon absorption in the bulk of the material resulting in an inter-band transition $|m\mathbf{k}\rangle \rightarrow |n\mathbf{k}\rangle$, with $\epsilon(n, \mathbf{k}) = \epsilon(m, \mathbf{k}) + \hbar\omega$. Here n, m are band indices and the

three-dimensional crystal momentum vector $\mathbf{k}$ is conserved up to a reciprocal lattice vector $\mathbf{G}$.

2. Propagation of the excited wave to the surface, with damping due to inelastic scattering.

3. Transmission through the surface by matching the Bloch wave $|n\mathbf{k}\rangle$ to a plane wave $\exp(i\mathbf{k}' \cdot \mathbf{r})$. The matching conditions are dictated by conservation of energy and the surface parallel component of $\mathbf{k}$, i.e. $\mathbf{k}'_{\|} = \mathbf{k}_{\|}$ and $k'^2_{\perp} = k^2_{\perp} + 2mV_0/\hbar^2$, where V_0 represents the surface potential barrier.

The three-step model is very useful for relating the photoemission data to the three-dimensional band structure of the material. However, for an accurate calculation of ARPES intensities, the one-step model should be used, where the photoelectron final state is calculated in all space (bulk, surface and vacuum) as a single wave function with proper boundary conditions (so-called 'time-reversed low-energy electron diffraction' boundary conditions). A suitable computational scheme is the layered Korringa–Kohn–Rostoker method [16].

The hole left behind in the photoemission process is not an independent particle, but it interacts with the electrons and the lattice, giving rise to various many-body effects, which are conveniently described using quasiparticle theory.

3.7 Quasiparticle Theory

In a photoemission experiment, an electron is ejected from the system, which becomes ionized. Neglecting the interaction between the photoelectron and the hole left behind, i.e. applying the so-called sudden approximation, the photoemission excitation is a one-electron removal process from the N-particle ground state to a $N - 1$ particle excited state. In the same fashion, inverse photoemission probes the one-electron addition process from the N-particle ground state to a $N + 1$-particle excited state. The true excitations are called quasiparticles. In the limit of vanishing electron interaction, the quasiparticle wave functions are the spin-orbitals of the ground state Slater determinant and the quasiparticle energies are the one-electron levels. In the independent particle approximation, the quasiparticles are taken as HF or Kohn–Sham orbitals. This neglects electron correlation and the interaction of the electrons with the lattice vibrations. These effects change the quasiparticle energies and wave functions. The quasiparticles are said to be renormalized or 'dressed' by the interaction. In particular, due to inelastic scattering at collective excitations such as phonons and plasmons, the one-electron quasiparticles will decay after a characteristic lifetime. As a result, compared to the delta-function-like photoemission peaks corresponding to the independent particle approximation, the true photoemission peaks are energy shifted and lifetime broadened. Moreover, some spectral weight of the main peak is lost to extra ('satellite') peaks, corresponding to some inelastic process.

3.7.1 Green's Functions

Quasiparticles can be described using many-body Green's function techniques. We introduce the retarded one-electron Green's function

$$G(\mathbf{r}, \mathbf{r}', t - t') = -i\theta(t - t')\langle 0|\{\Psi(\mathbf{r}t), \Psi^+(\mathbf{r}'t')\}|0\rangle, \qquad (3.33)$$

where $\Psi^+(\mathbf{r}t)$ is a Heisenberg field operator which creates an electron at point $\mathbf{r}$ and time t, and Ψ destroys one. $|0\rangle$ is the many-particle ground state and $\{A, B\} = AB + BA$ denotes the anti-commutator. Note that we have suppressed spin for convenience. This Green's function, or 'propagator', gives the probability amplitude for an electron to be found at $\mathbf{r}t$ if one was added at $\mathbf{r}'t'$. The one-electron removal and addition spectrum is given by the spectral function

$$A(\mathbf{k}, \omega) = -\frac{1}{\pi} \operatorname{Im} G(\mathbf{k}, \omega), \qquad (3.34)$$

where $G(\mathbf{k}, \omega)$ is the space and time Fourier transform of (3.33). In the following, we focus on a perfect crystal and suppress the band index. The Hamiltonian of the non-interacting system is then given by $H_0 = \sum_{\mathbf{k}} \epsilon_{\mathbf{k}} \hat{n}_{\mathbf{k}}$, where $\mathbf{k}$ labels the Bloch eigenstates with energy $\epsilon_{\mathbf{k}}$ and $\hat{n}_{\mathbf{k}}$ is the corresponding occupation number operator. It is easy to see that in this non-interacting case, Green's and spectral functions are given by

$$G_0(\mathbf{k}, \omega) = (\omega - \epsilon_{\mathbf{k}} + i\eta)^{-1}, \quad A(\mathbf{k}, \omega) = \delta(\omega - \epsilon_{\mathbf{k}}). \qquad (3.35)$$

Thus, the photoemission peaks are delta functions, meaning that Bloch states are exact excitations of energy $\epsilon_{\mathbf{k}}$ (band energy) and infinite lifetime. As mentioned above, due to electron interaction, the true photoemission peaks are shifted, broadened and may have satellite structures. In quasiparticle theory, these effects are described by the so-called self-energy Σ, which is essentially the difference between the inverses of the exact and the free Green's function. The self-energy is defined through the Dyson equation

$$G = G_0 + G_0 \Sigma G \quad \Leftrightarrow \quad G^{-1} = G_0^{-1} - \Sigma. \qquad (3.36)$$

For a single band in a crystal, we have

$$G^{-1}(\mathbf{k}, \omega) = \omega - \epsilon_{\mathbf{k}} - \Sigma_{\mathbf{k}}(\omega). \qquad (3.37)$$

It is clear from (3.35) and (3.37) that $\operatorname{Re}\Sigma$ describes a shift of the eigenvalues $\epsilon_{\mathbf{k}}$ (band energy) and $\operatorname{Im}\Sigma$ results in peak broadening, i.e. it reflects the finite lifetime $\tau = \hbar/\operatorname{Im}\Sigma$ of the quasiparticle. There are various methods to find (approximate) self-energies. For the electron correlation effect, two of the most popular methods are the so-called GW approximation and dynamical mean-field theory.

3.7.2   GW Approximation

The GW approximation was invented by Hedin in 1965 [17] and owes its name from the form of this self-energy, which is $\Sigma = iGW$, i.e. the product (or convolution) of Green's function (G) and the screened Coulomb interaction (W). The latter is given by [18]

$$W(\mathbf{r}, \mathbf{r}', \omega) = \frac{e^2}{4\pi\epsilon_0} \int d\mathbf{r}'' \frac{\epsilon^{-1}(\mathbf{r}, \mathbf{r}'', \omega)}{|\mathbf{r} - \mathbf{r}'|} , \qquad (3.38)$$

where $\epsilon(\mathbf{r}, \mathbf{r}', \omega)^{-1}$ is the inverse dielectric function. This expression may be understood by analogy to the electrostatic energy between two electrons in a polarizable medium, which is given by $e^2/[4\pi\epsilon_0\epsilon_r |\mathbf{r} - \mathbf{r}'|]$, where ϵ_r is the relative dielectric permittivity. The dielectric function $\epsilon(\mathbf{r}, \mathbf{r}', \omega)$ generalizes ϵ_r to inhomogeneous media and dynamic screening effects. The GW approximation is most often used in a non-self-consistent way, i.e. as $\Sigma = iG_0 W_0$ with the free Green's function G_0 instead of the full Green's function G. The GW approximation has been very successful for correcting band energies of weakly correlated systems. In particular, bandgaps of semi-conductors are very well reproduced in the GW approximation, while the values obtained in DFT (except for DFT-HF hybrid functionals) are systematically too small [19].

3.7.3   Bethe–Salpeter Equation

At this point, we briefly switch back to the problem of absorption spectroscopy. Since light absorption creates an electron–hole pair, absorption spectra are described with an electron–hole (i.e. a two-particle) Green's function G_{eh}. If the excited electron and the hole do not interact, G_{eh} is just the product of the one-particle removal (hole) Green's function G_h and the addition (electron) Green's function G_e. Electron–hole interaction leads to coupling of these two Green's functions, which may be expressed in a Dyson-type equation as [20]

$$G_{eh}(1, 2; 1', 2') = G_e(1, 1')G_h(2, 2') \qquad (3.39)$$
$$+ \int G_e(1, 3)G_h(2, 4)K(3, 4; 5, 6)G_{eh}(5, 6; 1', 2')d3d4d5d6 ,$$

where 1 stands for all coordinates of particle 1 and K is the interaction kernel. In the Bethe–Salpeter equation (BSE) approach, (3.39) is solved with K given by the screened Coulomb interaction in (3.38) and the bare exchange interaction. The electron and hole Green's functions, G_e and G_h, are commonly computed in the GW approximation. The BSE approach is arguably the most accurate first-principles method for absorption spectroscopy in solids, but it is computationally very demanding. It was first applied to X-ray spectra by Shirley in 1998 [21]. It accounts well for strong excitonic effects and features electron–hole multiplet coupling in L-edge spectra [22]. Let us note that the latter effect is also well described with multichannel multiple scattering theory [23, 24], where the electron–hole coupling is dealt by a CI calculation of the scattering matrix. However, at present, none of these particle–hole theories can fully account for the complex multiplet structure of L-edge spectra of open-shell transition metal compounds. These spectra are still best described with CI methods, either the semi-empirical ligand-field multiplet model (see Chap. 4) or the ab initio complete active space approach on small clusters [25].

3.7.4 Static and Dynamical Mean-Field Theory

In strongly correlated electron systems, e.g. $3d$ transition metal oxides and $5f$ elements, collective phenomena such as band magnetism, metal–insulator transition and high-T_c superconductivity are observed. These are genuine many-body effects that cannot be explained in the independent particle picture. Itinerant magnetism and the metal–insulator transition are due to the competition between the kinetic energy, which leads to delocalized band states, and strong local Coulomb repulsion which favours electron localization and formation of magnetic moments. The most simple model to study these problems is the (one-band) Hubbard model [26], whose Hamiltonian is given by

$$H = \sum_{k\sigma} \epsilon_k n_{k\sigma} + U \sum_i n_{i\uparrow} n_{i\downarrow} = \sum_{ij\sigma} t_{ij} c_{i\sigma}^+ c_{j\sigma} + U \sum_i n_{i\uparrow} n_{i\downarrow},$$

where lattice sites are labelled by i and j, the crystal momentum by k and spin by σ. Further, c_ν^+ (c_ν) creates (destroys) an electron in state ν, and $n_\nu \equiv c_\nu^+ c_\nu$ counts them. t_{ij} is the hopping (or 'transfer') integral between sites i and j, and U is the Coulomb energy between two electrons occupying the same site. The corresponding one-electron Green's function is

$$G_{k\sigma}(\omega) = [\omega - \epsilon_k - \Sigma_{k\sigma}(\omega)]^{-1} .$$

Despite the apparent simplicity of the Hubbard model, the exact solution is unknown (except in one dimension) and the self-energy Σ must be approximated. At the lowest level, there is the normal mean-field (i.e. HF) approximation, where Σ is taken to be static, i.e. independent of frequency ω. It is given by $\Sigma_{k\sigma} = U\langle n_{k-\sigma}\rangle$, where the occupation numbers $\langle n_{k-\sigma}\rangle$ must be calculated self-consistently. When the lowest energy solution corresponds to a different occupation between spin-up and spin-down bands $\langle n_{k\uparrow}\rangle \neq \langle n_{k\downarrow}\rangle$, the band energies $E_{k\sigma} = \epsilon_k + U\langle n_{k-\sigma}\rangle$ become exchange split, and the ground state is ferromagnetic. HF and LDA are such static mean-field theories and can account for certain static exchange effects, such as ferromagnetism. But they lack all dynamic correlation, which is crucial for the metal–insulator transition and for various phenomena seen in photoemission spectra, such as band narrowing and satellite structures.

Dynamic correlation effects can, to some extend, be described by the *dynamical mean-field theory* (DMFT) [27], where the self-energy is taken to be frequency dependent but local, i.e. momentum independent, $\Sigma_{k\sigma}(\omega) \to \Sigma_\sigma(\omega)$. We note that self-energies from static mean-field theory (such as LDA) and dynamical mean-field theory can be, and often are, combined. Although the total self-energy is then both momentum and frequency dependent, it is still an approximation.

The basic idea of DMFT is to map the Hubbard model with correlation ($U \neq 0$) on all lattice sites onto the Anderson model, which describes one correlated atom (the 'impurity') coupled to an effective bath of band states. In the Anderson model, we have $U \neq 0$ only at the impurity site ($i = 0$), and as a consequence, the self-energy is a frequency dependent, but local, quantity $\Sigma_{ij}(\omega) = \Sigma_0(\omega)\delta_{i0}\delta_{ij}$. In DMFT, this local self-energy is taken as the self-energy of the lattice problem (Hubbard model). The mapping, i.e. the definition of the effective bath, must be done in a self-consistent manner such that the on-site matrix elements of the lattice model Green's function $G_{ii}(\omega)$ coincide with those of the impurity model [27]. While the Anderson impurity model is simpler than the Hubbard model, it is nonetheless a complex many-body problem. Implementations of DMFT mostly differ in the approximations used for solving the impurity problem.

DMFT has been applied to photoemission spectroscopy of correlated systems [28] and results in improved spectra compared to independent particle approximation (HF or LDA). In transition metal systems, for example, photoemission spectra calculated in DMFT can account for finite temperature effects, correlation-driven band narrowing and satellite peaks [29].

3.8   Conclusions

In this chapter, I have tried to give a brief introduction to the theory of X-ray absorption and photoemission spectroscopy. Along the way, it appeared useful to present succinctly the principles of several computational methods of electronic structure that are used in spectroscopic calculations. Given the vast nature of the subject, this account is necessarily very incomplete. But I hope that the reader got an idea of the physics underlying the different theoretical methods and that it aroused his/her curiosity to dwell deeper into the subject by reading some of the cited literature.

References

1. L.I. Schiff, *Quantum Mechanics*, 3rd edn. (McGraw-Hill, New York, 1968)
2. R.D. Cowan, *The Theory of Atomic Structure and Spectra* (University of California Press, Berkeley, 1981)

3. V.A. Fock, Näherungsmethode zur Lösung des quantenmechanischen Mehrkörperproblems. Z. Phys. **61**, 126 (1930). https://doi.org/10.1007/BF01340294

4. P. Hohenberg, W. Kohn, Inhomogeneous electron gas. Phys. Rev. **136**, B864 (1964). https://doi.org/10.1103/PhysRev.136.B864

5. W. Kohn, L.J. Sham, Self-consistent equations including exchange and correlation effects. Phys. Rev. **140**, A1133 (1965). https://doi.org/10.1103/PhysRev.140.A1133

6. O. Bunău, M. Calandra, Projector augmented wave calculation of X-ray absorption spectra at the $L_{2,3}$ edges. Phys. Rev. B **87**, 205105 (2013). https://doi.org/10.1103/PhysRevB.87.205105

7. P. Krüger, M. Sluban, P. Umek, P. Guttman, C. Bittencourt, Chemical bond modification upon phase transformation of TiO_2 nanoribbons revealed by nanoscale X-ray linear dichroism. J. Phys. Chem. C **121**, 17038 (2017). https://doi.org/10.1021/acs.jpcc.7b06968

8. A. Zangwill, P. Soven, Density-functional approach to local field effects in finite systems: photoabsorption in the rare gases. Phys. Rev. A **21**, 1561 (1980). https://doi.org/10.1103/PhysRevA.21.1561

9. E. Runge, E.K.U. Gross, Density-functional theory for time-dependent systems. Phys. Rev. Lett. **52**, 997 (1984). https://doi.org/10.1103/PhysRevLett.52.997

10. J. Schwitalla, H. Ebert, Electron core-hole interaction in the X-ray absorption spectroscopy of $3d$ transition metals. Phys. Rev. Lett. **80**, 4586 (1998). https://doi.org/10.1103/PhysRevLett.80.4586

11. M. Stener, G. Fronzoni, M. de Simone, Time dependent density functional theory of core electrons excitations. Chem. Phys. Lett. **373**, 115 (2003). https://doi.org/10.1016/S0009-2614(03)00543-8

12. O. Bunău, Y. Joly, Time-dependent density functional theory applied to X-ray absorption spectroscopy. Phys. Rev. B **85**, 155121 (2012). https://doi.org/10.1103/PhysRevB.85.155121

13. A.L. Ankudinov, A.I. Nesvizhskii, J.J. Rehr, Dynamic screening effects in X-ray absorption spectra. Phys. Rev. B **67**, 115120 (2003). https://doi.org/10.1103/PhysRevB.67.115120

14. C. Westphal, The study of the local atomic structure by means of X-ray photoelectron diffraction. Surf. Sci. Rep. **50**, 1 (2003). https://doi.org/10.1016/S0167-5729(03)00022-0

15. C.N. Berglund, W.E. Spicer, Photoemission studies of copper and silver: theory. Phys. Rev. **136**, A1030 (1964). https://doi.org/10.1103/PhysRev.136.A1030

16. J.B. Pendry, Theory of photoemission. Surf. Sci. **57**, 679 (1976). https://doi.org/10.1016/0039-6028(76)90355-1

17. L. Hedin, New method for calculating the one-particle Green's function with application to the electron-gas problem. Phys. Rev. **139**, A796 (1965). https://doi.org/10.1103/PhysRev.139.A796

18. F. Aryasetiawan, O. Gunnarsson, The GW method. Rep. Prog. Phys. **61**, 237 (1998). https://doi.org/10.1088/0034-4885/61/3/002

19. M. van Schilfgaarde, T. Kotani, S. Faleev, Quasiparticle self-consistent GW theory. Phys. Rev. Lett. **96**, 226402 (2006). https://doi.org/10.1103/PhysRevLett.96.226402

20. G. Onida, L. Reining, A. Rubio, Electronic excitations: density-functional versus many-body Green's-function approaches. Rev. Mod. Phys. **74**, 601 (2002). https://doi.org/10.1103/RevModPhys.74.601

21. E.L. Shirley, Ab initio inclusion of electron-hole attraction: application to X-ray absorption and resonant inelastic X-ray scattering. Phys. Rev. Lett. **80**, 794 (1998). https://doi.org/10.1103/PhysRevLett.80.794

22. E.L. Shirley, Bethe–Salpeter treatment of X-ray absorption including core-hole multiplet effects. J. Electron. Spectrosc. Relat. Phenom. **144–147**, 1187 (2005). https://doi.org/10.1016/j.elspec.2005.01.191

23. P. Krüger, C.R. Natoli, X-ray absorption spectra at the Ca $L_{2,3}$ edge calculated within multichannel multiple scattering theory. Phys. Rev. B **70**, 245120 (2004). https://doi.org/10.1103/PhysRevB.70.245120

24. P. Krüger, Multichannel multiple scattering calculation of $L_{2,3}$-edge spectra of TiO_2 and $SrTiO_3$: importance of multiplet coupling and band structure. Phys. Rev. B **81**, 125121 (2010). https://doi.org/10.1103/PhysRevB.81.125121

25. K. Ogasawara, T. Iwata, Y. Koyama, T. Ishii, I. Tanaka, H. Adachi, Relativistic cluster calculation of ligand-field multiplet effects on cation $L_{2,3}$ X-ray-absorption edges of $SrTiO_3$, NiO, and CaF_2. Phys. Rev. B **64**, 115413 (2001). https://doi.org/10.1103/PhysRevB.64.115413

26. J. Hubbard, Electron correlations in narrow energy bands. Proc. R. Soc. London Ser. A **276**, 238 (1963). https://doi.org/10.1098/rspa.1963.0204

27. A. Georges, G. Kotliar, W. Krauth, M.J. Rozenberg, Dynamical mean-field theory of strongly correlated fermion systems and the limit of infinite dimensions. Rev. Mod. Phys. **68**, 13 (1996). https://doi.org/10.1103/RevModPhys.68.13

28. J. Braun, J. Minár, H. Ebert, M.I. Katsnelson, A.I. Lichtenstein, Spectral function of ferromagnetic $3d$ metals: a self-consistent LSDA + DMFT approach combined with the one-step model of photoemission. Phys. Rev. Lett. **97**, 227601 (2006). https://doi.org/10.1103/PhysRevLett.97.227601

29. M.I. Katsnelson, A.I. Lichtenstein, Electronic structure and magnetic properties of correlated metals. Eur. Phys. J. B **30**, 9 (2002). https://doi.org/10.1140/epjb/e2002-00352-1

4

X-ray Dichroisms in XAS

**Hebatalla Elnaggar, Pieter Glatzel, Marius Retegan, Christian Brouder
and Amélie Juhin**

Abstract In this book chapter, our goal is to provide experimentalists and theoreticians with an accessible approach to the measurement or calculation of X-ray dichroisms in X-ray absorption spectroscopy (XAS). We start by presenting the key ideas of different calculation methods such as density functional theory (DFT) and ligand-field multiplet (LFM) theory and discuss the pros and cons for each approach. The second part of the chapter is dedicated to the expansion of the XAS cross section using spherical tensors for electric dipole and quadrupole transitions. This expansion enables to identify a set of linearly independent spectra that represent the smallest number of measurements (or calculations) to be performed on a sample, in order to extract all spectroscopic information. Examples of the different dichroic effects which can be expected depending on the type of transitions and on the symmetry of the system are then given.

4.1 Introduction

4.1.1 The X-ray Absorption Cross Section

The X-ray absorption cross section is obtained by dividing the transition rate by the flux of photons and summing over all possible final states. It is given in (4.1) where $\hbar\omega$ is the photon energy, and α the fine structure constant. $I(E_I)$ and $F(E_F)$ are the initial and final state wave functions (energies), and T is the transition operator,

H. Elnaggar
Debye Institute for Nanomaterials Science, Utrecht University, 3584 CA Utrecht,
The Netherlands

P. Glatzel · M. Retegan
European Synchrotron Radiation Facility, 71 Rue des Martyrs, 38000 Grenoble, France

Ch. Brouder · A. Juhin (✉)
Institut de Minéralogie, Physique des Matériaux et Cosmochimie, CNRS-Sorbonne Université,
4 Place Jussieu, 75252 Paris Cedex 05, France
e-mail: amelie.juhin@sorbonne-universite.fr

$$\sigma_\omega = 4\pi^2 \alpha \hbar \omega \sum_F \langle I|T^\dagger|F\rangle \langle F|T|I\rangle \delta(E_I + \hbar\omega - E_F) \,. \tag{4.1}$$

The transition operator describes the interaction of photons with the system. In the case of an electromagnetic plane wave, the transition operator writes $T \propto e^{i\mathbf{k}\cdot\mathbf{r}} \left[\hbar\boldsymbol{\epsilon} \cdot \boldsymbol{\nabla} - \frac{g}{2s}\mathbf{k} \times \boldsymbol{\epsilon}\right]$ where $\boldsymbol{\epsilon}$ is the polarization vector of the incident photon, and $\mathbf{k}$ is the incident wave vector, g the gyromagnetic ratio ($g \approx 2$ for the electron), and s the electron spin [1]. The exponential in the transition operator can be expanded as a Taylor series

$$e^{i\mathbf{k}\cdot\mathbf{r}} \approx 1 + i\mathbf{k}\cdot\mathbf{r} - \frac{(\mathbf{k}\cdot\mathbf{r})^2}{2!} + \dots \tag{4.2}$$

The first term in the expansion approximates the interaction of the light with the atom as an electric dipole ($\langle F|T|I\rangle = \hbar\langle F|\boldsymbol{\epsilon}\cdot\boldsymbol{\nabla}|I\rangle = -\frac{m(E_F-E_I)}{\hbar}\langle F|\boldsymbol{\epsilon}\cdot\mathbf{r}|I\rangle$). The second term gives rise to the electric quadrupole interaction ($-i\frac{m(E_F-E_I)}{\hbar}\langle F|\boldsymbol{\epsilon}\cdot\mathbf{r}\mathbf{k}\cdot\mathbf{r}|I\rangle$) and to the (negligible) magnetic dipole one ($-\frac{1}{2}\langle F|(\boldsymbol{\epsilon} \times \mathbf{k}) \cdot (\mathbf{L} + g\mathbf{s})|I\rangle$) [1], the third term is the octupole transition ($\frac{m(E_F-E_I)}{6\hbar}\langle F|(\boldsymbol{\epsilon}\cdot\mathbf{r})(\mathbf{k}\cdot\mathbf{r})^2|I\rangle$) [2], and so on. In this chapter, we focus on electric dipole and quadrupole transitions.

The summation over final states in (4.1) implies that one has first to calculate the ground state, all possible final states, and then compute the transition matrix elements between the ground state and the final states. This is not always the most efficient way to numerically calculate XAS. Instead, Green's function can be used to replace the summation over final states by a propagator of the transition operator. Hence $\sum_F |F\rangle\langle F|\delta(E_I + \hbar\omega - E_F) \to \frac{-1}{2\pi i}(G^+ - G^-)$ with $G^\pm(E_I + \hbar\omega) = \frac{1}{\hbar\omega - H_F + E_I \pm \frac{1}{2}i\Gamma}$, where H_F is the final state Hamiltonian. The "Fermi Golden Rule" can be expressed as in (4.3). Most modern codes calculating core level spectra use this expression

$$\sigma_\omega = -4\pi\alpha\hbar\omega\,\mathrm{Im}\left[\langle I|T^\dagger G^+(E_I + \hbar\omega)T|I\rangle\right] \,. \tag{4.3}$$

Let us now discuss electric dipole transitions according to the first term of the expansion in (4.2). For electric dipole transitions we have $T = \boldsymbol{\epsilon}\cdot\mathbf{r}$. One can see from the expression of the transition operator that the cross section will depend on the orientation of the polarization vector ($\boldsymbol{\epsilon}$) with respect to the absorbing system ($\mathbf{r}$). The X-ray absorption spectrum measured on any sample is in fact the sum of several linearly independent spectra as will be discussed further in this chapter. They can be disentangled by macroscopically orienting the sample, e.g., by using a single crystal or orienting the magnetic moments. Consequently, one may wonder:

- How many independent spectra exist for a given system?
- What information do they give us about the absorbing system?

4.1.2 Definition of Dichroisms

X-ray dichroism can be defined as the difference in the X-ray absorption cross section measured for two orthogonal polarization states of the incident light. There exist different types of dichroism. Dichroism measurements can be classified according to the type of polarization used for the measurements into linear and circular. *Linear dichroism* (LD) is the difference measured with linearly polarized light, where in most cases the polarization vector is set parallel and perpendicular to an orientation axis, while *circular dichroism* (CD) is the difference measured with circularly polarized light (left handed and right handed).

Not all systems exhibit dichroism effects when the polarization of the light is changed. Certain symmetry conditions regarding the interaction operator between light and matter have to be satisfied for dichroism effects to occur, which brings us to the second classification of dichroism types. Two symmetry operations are essential for this classification:

- Time-reversal symmetry,
- Space inversion (also called parity).

Natural dichroism (ND) refers to dichroism effects that occur in non-magnetic systems where time-reversal symmetry is conserved (i.e., the system is even under time-reversal operation). Using linearly polarized light, one can measure *X-ray natural linear dichroism* (XNLD). On the other hand, using circularly polarized light, one can measure *X-ray natural circular dichroism* (XNCD) only for systems that do not have a centre of inversion (i.e., the system is of odd parity).

Magnetic dichroism (MD) relates to dichroism effects measured in magnetic (ferro, ferri, or antiferromagnetic) systems where time-reversal symmetry is broken either by spontaneous magnetic ordering in the sample or by the application of an external magnetic field. *X-ray magnetic linear dichroism* (XMLD) is parity-even and time-reversal even and *non-reciprocal linear dichroism* (NRLD) is parity-odd and time-reversal odd. Using circularly polarized light, *X-ray magnetic circular dichroism* (XMCD) and *X-ray magneto-optical dichroism* (XMχD) effects can be measured. The former is parity-even and time-reversal odd while the later is parity-odd and time-reversal odd.

In magnetic materials, which will be discussed in this chapter, several cases are possible:

- In the case of centrosymmetric crystals with ferro- or ferrimagnetic properties, one can measure XMCD.
- In the case of centrosymmetric crystals with antiferromagnetic properties, XMLD can be measured.
- In the case of magnetized, non-centrosymmetric crystals, XMχD and NRLD can be measured.

XMCD and XMLD measurements give, respectively, access to the average value of $\langle M \rangle$ and $\langle M^2 \rangle$ of the local magnetization for the absorber. On the other hand,

XMχD and NRLD signals are related to moments that are more complex, the anapole orbital moment and other higher order moments.

4.1.3   The Many-Body Problem in Spectra Calculations

The calculation of an absorption spectrum is a formidable task: it requires the calculation of the ground state of the system, the excited states of the system, and the interaction of the system with the electromagnetic field (X-ray beam). This means that the theoretical approach required to calculate XAS has to be suitable for calculating the electronic structure in addition to properly considering the interaction with the electromagnetic field. Approximations have often to be made to calculate the absorption (or the scattering Kramers–Heisenberg) cross section and the spectroscopist therefore has to choose which theoretical approach is the most suitable for the problem at hand.

In principle, the ground and excited states can be determined by solving the Dirac equation which accounts for all relativistic effects and includes all possible interactions in the Hamiltonian. This full treatment provides a relativistic, many-body, extended description of the electronic states. Unfortunately, in practice, it is not possible to perform such a calculation as it is computationally very consuming. In most cases one solves instead the Schrödinger equation and introduces relativistic effects as perturbations (e.g., the spin-orbit interaction). Furthermore, one can make use of the Born–Oppenheimer approximation to separate the electronic properties of the system from the dynamics of the nuclei. In order to describe the electronic part of the wave function, various theoretical approaches can be used such as (i) the single-particle extended picture (DFT-based approaches), (ii) the many-body atomic picture (multiplet theory), and (iii) the many-body extended picture (beyond DFT methods).

4.1.3.1   The Single-Particle Extended Picture of Electronic States

DFT-based methods can be used to describe the electronic states using a single-particle extended picture. Although DFT methods should formally only apply to ground state calculations, they are often used for the calculation of excited states probed in core level spectroscopies. DFT methods simplify the ground state wave function of N electrons by replacing them with a fictitious, non-interacting system of independent electrons, that have the same electronic density as the real system. The correct charge density minimizes the total energy of the system. The Schrödinger equation is transformed into a system of equations (called the "Kohn–Sham equations") with an effective Hamiltonian and wave functions, which are functions of only one space variable. This implies that DFT is essentially a single-particle approach, although some many-particle (many-body) interactions are contained in the exchange and correlation term of the electronic effective potential. The exact analytical expression of this term is not known. This means that DFT can only be applied in an approximate form for example using the local density approximation (LDA) or the generalized gradient approximation (GGA)].

One can determine the ground state wave function by solving the Kohn–Sham equations which are constructed from the single-particle wave functions. The ground state wave function is therefore formed by a single Slater determinant. This is an important point because it limits the ability to treat a many-body response of the system described through a linear combination of Slater determinants. To illustrate this let us take an example with two electrons. Coupling $s = 1/2$ to $s = 1/2$ yields $S = 1$ or 0. This corresponds to four $|S, Ms\rangle$ wave functions: $|S = 1, Ms = 1\rangle$, $|S = 1, Ms = -1\rangle$ $|S = 1, Ms = 0\rangle$, and $|S = 0, Ms = 0\rangle$. These functions need to meet the property of being anti-symmetric under particle exchange and it can be shown that only $|S = 1, Ms = 1\rangle$, $|S = 1, Ms = -1\rangle$ can each be expressed as a single Slater determinant. However, the $|S = 1, Ms = 0\rangle$ and $|S = 0, Ms = 0\rangle$ states can only be expressed as the combination of two Slater determinants and thus cannot be calculated in DFT.

One can group the various DFT-based methods according to their characteristics:

– Cluster or periodic: The Kohn–Sham equations can be solved either for a cluster centred around the absorbing atom (direct or real space methods) or starting from a unit cell of the crystal (or a multiple unit cell, which is called supercell) in order to take advantage of the 3D periodicity (reciprocal space methods).
– Self-consistency or not: The Kohn–Sham equations can be solved without or (preferably) with self-consistency, i.e., using an iterative cycle where two successive steps are mixed until a convergence criterion is reached to determine the charge density.
– Type of basis functions used to expand the orbital solutions of the Kohn–Sham equations: either localized functions [linear combination of atomic orbitals (LCAO), linear muffin-tin orbitals (LMTO)], or delocalized functions [plane waves (PW), full-potential linearized augmented plane waves (FLAPW)].
– Approximation made on the shape of the electronic potential: For example in LMTO or multiple scattering theory, the potential is approximated to be spherically symmetric in the atoms, and constant between them (muffin-tin). In full-potential methods [FLAPW, or projector augmented wave (PAW)-pseudopotentials], no approximation is made, which is generally preferable, even though it makes the calculations more consuming.

4.1.3.2 The Many-Body Atomic Picture of Electronic States

A Simple Introduction to the Many-Body Atomic Picture

Let us consider as an example a Cr^{3+} ion in an octahedral (O_h) environment. Here the solid is reduced to an atom embedded in a mean field known as the crystal field (CF) that mimics the effect of the inter-atomic interactions. The atomic electronic configuration is $1s^2 2s^2 2p^6 3s^2 3p^6 3d^3$. The degeneracy of the Cr $3d$ levels is lifted due to the CF and the $3d$ orbitals are split into two groups: the e_g orbitals pointing towards the ligands and the t_{2g} orbitals pointing between the ligands. The number of possible electronic states is given by the number of allowed arrangements of three electrons into ten spin orbitals, i.e., $C_{10}^3 = 120$ microstates. The energy separation between these states arises due to the combined effect of: (i) CF splitting, (ii) electronic repulsions, and (iii) spin-orbit coupling, i.e., the multiplet effects. Electrons occupying closed shells do not actually contribute to the energy splitting of the electronic levels; there is only one way to completely fill a shell giving a single average energy of the configuration.

All these multiplet states can be further grouped in so-called *term symbols* (or spectroscopic terms) according to their energy, spin and orbital moments. The relative energy positions of these spectroscopic terms for a $3d^n$ transition metal ion in O_h CF (and neglecting $3d$ spin-orbit coupling) were calculated by Tanabe and Sugano and are available in several references (e.g., [3, 4]). Similar diagrams are available in [5] for symmetries lower than O_h, such as trigonal or tetragonal. The relative energies of the electronic states depend on the CF parameters as well as the Racah parameters that relate to the electronic repulsions. The determination of the spectroscopic terms becomes very complex when the spin-orbit coupling and Zeeman terms are included in the Hamiltonian and/or if lower symmetries are considered. LFM theory takes these effects into account and has been realized in several computer codes.

Key Ideas of Ligand-Field Multiplet Theory

Atomic multiplet theory, crystal field multiplet theory, and LFM theory (sometimes collectively referred to as the *multiplet theory*) are based on concepts that were developed in atomic physics and make use of group theory. One has to solve the Schrödinger equation for the ion with its N electrons in a given configuration

$$\hat{H}|g\rangle = E|g\rangle , \tag{4.4}$$

where $\hat{H}$ is the Hamiltonian of the system for the chosen configuration, E and $|g\rangle$ are the eigenvalue and eigenstate, respectively. The different eigenstates are functions of N electrons, hence they are called many-body (or multi-electronic) states. The Hamiltonian is expressed as

$$\hat{H} = \hat{T} + \hat{V} + \hat{V}_{ee} + \hat{H}_{SO} + \hat{H}_{CF} , \tag{4.5}$$

where $\hat{T}$ is the kinetic energy of the electrons, $\hat{V}$ the Coulomb attraction between electrons and the nucleus, $\hat{V}_{ee}$ the electron–electron Coulomb interaction, $\hat{H}_{SO}$ the spin-orbit coupling interaction, and $\hat{H}_{CF}$ the CF Hamiltonian, which takes into account the local environment of the absorbing atom.

These interactions will now be expressed in second quantization formalism. In this notation, any operator can be expressed in terms of creation ($c_\tau^\dagger$) and annihilation (c_τ) operators. The operator $c_\tau^\dagger$ creates a state characterized by the quantum numbers τ (for example, if we choose to express the states as spin-orbitals, τ will be the set of quantum numbers n, l, m, σ that give the principal quantum number, the orbital momentum, the projected orbital momentum, and the projected spin momentum that uniquely identify this state) when it acts on the vacuum state $|0\rangle$. The operator c_τ is the annihilation operator of the state τ. In this formalism, the (spherical) atomic interactions write

$$\hat{T} = \sum_{\tau_1, \tau_2} \langle \tau_1 | \frac{\hat{p}^2}{2m} | \tau_2 \rangle c^\dagger_{\tau_1} c_{\tau_2} \,, \tag{4.6}$$

$$\hat{V} = \sum_{\tau_1, \tau_2} \langle \tau_1 | - \frac{Ze^2}{\hat{r}} | \tau_2 \rangle c^\dagger_{\tau_1} c_{\tau_2} \,, \tag{4.7}$$

$$\hat{V}_{ee} = \frac{1}{2} \sum_{\tau_1, \tau_2, \tau_3, \tau_4} \langle \tau_1 \tau_2 | \frac{e^2}{|\hat{r} - \hat{r}'|} | \tau_3 \tau_4 \rangle c^\dagger_{\tau_2} c^\dagger_{\tau_1} c_{\tau_3} c_{\tau_4} \,, \tag{4.8}$$

$$\hat{V}_{SO} = \sum_{\tau_1, \tau_2} \langle \tau_1 | \xi \hat{l}.\hat{s} | \tau_2 \rangle c^\dagger_{\tau_1} c_{\tau_2} \,. \tag{4.9}$$

Here $\hat{p}$ is the linear momentum operator, m is the electron mass, e is the electron charge, $\hat{r}$ is the position operator, $\hat{l}$ and $\hat{s}$ are the orbital and spin momenta operators, and ξ is an atom dependent constant that is a function of the gradient of the atomic potential ($\xi \propto \frac{1}{r} \frac{dV}{dr}$). The kinetic energy of the electrons and the Coulomb interaction of the electrons with the nucleus are fixed for a given atomic configuration and they contribute only to the average energy of the configuration; hence $\hat{T}$ and $\hat{V}$ do not contribute to the multiplet splitting, and will not be further discussed. As a matter of fact, they are typically not evaluated in standard multiplet calculation programs. However, we are left with the task of simplifying the terms $\hat{V}_{SO}$ and $\hat{V}_{ee}$. Let us start with $\hat{V}_{SO}$ and assume that z is the quantization axis.

Influence of Spin-Orbit Coupling Interaction

The spin-orbit interaction is given as follows:

$$\hat{V}_{SO} = \sum_{\tau_1, \tau_2} \langle \tau_1 | \xi (l_x s_x + l_y s_y + l_z s_z) | \tau_2 \rangle c^\dagger_{\tau_1} c_{\tau_2}$$

$$= \sum_{\tau_1, \tau_2} \langle \tau_1 | \xi l_z s_z | \tau_2 \rangle c^\dagger_{\tau_1} c_{\tau_2} \tag{4.10}$$

$$+ \sum_{\tau_1, \tau_2} \langle \tau_1 | \xi (l_x s_x + l_y s_y) | \tau_2 \rangle c^\dagger_{\tau_1} c_{\tau_2} \,.$$

We shall use a set of atomic spin-orbitals as basis functions, $\psi_i = R_{n_i, l_i}(r) Y_{l_i, m_i}(\theta, \phi) \sigma_i$ where $Y_{l_i, m_i}(\theta, \phi)$ is the spherical harmonic, and $R_{n_i, l_i}(r)$ is the radial part. Given that the potential in ξ has a spherical form one can separate the radial and angular parts of the Hamiltonian. The angular part of the first term gives

$$\sum_{(l_1,m_1),(l_2,m_2)} \langle Y_{l_1,m_1}\sigma_1 | l_z s_z | Y_{l_2,m_2}\sigma_2 \rangle c^{\dagger}_{l_1,m_1,\sigma_1} c_{l_2,m_2,\sigma_2}$$

$$= \frac{1}{2} \sum_{m=-l}^{l} m(c^{\dagger}_{l,m,\uparrow} c_{l,m,\uparrow} - c^{\dagger}_{l,m,\downarrow} c_{l,m,\downarrow}) \,. \tag{4.11}$$

The second term gives

$$\sum_{(l_1,m_1),(l_2,m_2)} \langle Y_{l_1,m_1}\sigma_1 | (l_x s_x + l_y s_y) | Y_{l_2,m_2}\sigma_2 \rangle c^{\dagger}_{l_1,m_1,\sigma_1} c_{l_2,m_2,\sigma_2}$$

$$= \sum_{(l_1,m_1),(l_2,m_2)} \langle Y_{l_1,m_1}\sigma_1 | \frac{1}{2}(l^+ s^- + l^- s^+) | Y_{l_2,m_2}\sigma_2 \rangle c^{\dagger}_{l_1,m_1,\sigma_1} c_{l_2,m_2,\sigma_2} \tag{4.12}$$

$$= \frac{1}{2} \sum_{m=-l}^{l-1} \sqrt{(l-m)(l+m+1)}(c^{\dagger}_{l,m+1,\downarrow} c_{l,m,\uparrow} + c^{\dagger}_{l,m,\uparrow} c_{l,m+1,\downarrow}) \,.$$

Hence the angular part of the spin-orbit Hamiltonian finally writes

$$\hat{H}_{SO} = \frac{1}{2} \sum_{m=-l}^{l} m(c^{\dagger}_{l,m,\uparrow} c_{l,m,\uparrow} - c^{\dagger}_{l,m,\downarrow} c_{l,m,\downarrow})$$

$$+ \frac{1}{2} \sum_{m=-l}^{l-1} \sqrt{(l-m)(l+m+1)}(c^{\dagger}_{l,m+1,\downarrow} c_{l,m,\uparrow} + c^{\dagger}_{l,m,\uparrow} c_{l,m+1,\downarrow}) \,. \tag{4.13}$$

It is clear from (4.13) that the spin-orbit interaction mixes states with different projected orbital and spin momenta.

The Electron–Electron Coulomb Interaction

Now we undertake the simplification of the electron–electron Coulomb interaction. This is more involved than the simplification of the spin-orbit coupling Hamiltonian. Cowan nicely explains the details of the derivation in his book [6]. We will rely on a combination of the derivations by Cowan [6] and Haverkort [7] in this section. This is not a thorough derivation; it is only meant to qualitatively explain the origin of multiplet splittings.

The first step to simplify this Hamiltonian is to perform a multipole expansion of the term $\frac{1}{\vec{r}-\vec{r'}} = \sum_{k=0}^{\infty} \sum_{m=-k}^{k} Y^*_{k,m}(\theta',\phi') \frac{4\pi}{2k+1} \frac{r^k_<}{r^{k+1}_>} Y_{k,m}(\theta,\phi)$, where $r^k_<$ and $r^k_>$ are, respectively, the lesser and greater of the distances r and r'. Now we are in a position to separate the radial and angular terms of the expression and separate the angular variables of each electron. Using the atomic spin-orbital basis to express the matrix elements of the angular part of the Hamiltonian one finds that

$$\sum_{\tau_1,\tau_2,\tau_3,\tau_4} \langle Y_{l_1,m_1}(\theta,\phi)\sigma_1 Y_{l_2,m_2}(\theta',\phi')\sigma_2| \sum_{k=0}^{\infty} \sum_{m=-k}^{k} Y_{k,m}^*(\theta',\phi')Y_{k,m}(\theta,\phi)|$$

$$Y_{l_3,m_3}(\theta,\phi)\sigma_3 Y_{l_4,m_4}(\theta',\phi')\sigma_4\rangle \qquad (4.14)$$

$$= \sum_{k=0}^{\infty} \sum_{m=-k}^{k} \sum_{\tau_1,\tau_2,\tau_3,\tau_4} \langle Y_{l_1,m_1}(\theta,\phi)\sigma_1|Y_{k,m}(\theta,\phi)|Y_{l_3,m_3}(\theta,\phi)\sigma_3\rangle$$

$$\langle Y_{l_2,m_2}(\theta',\phi')\sigma_2|Y_{k,m}^*(\theta',\phi')|Y_{l_4,m_4}(\theta',\phi')\sigma_4\rangle .$$

We have in (4.14) integrals involving three spherical harmonics which are given by the Gaunt coefficients. This can be used to restrict the values of the summation over k and m. The Gaunt coefficients are different from zero in the first integral only for $m = m_1 - m_3$. Similarly, the second integral is different from zero for $m = m_4 - m_2$. Hence, in combination, one concludes that the total M_z is conserved ($m_1 + m_2 = m_3 + m_4$) for the integrals. Furthermore, the values of k are restricted to values of $k \leq min(|l_1 + l_4|, |l_2 + l_3|)$. This simplifies the angular part.

Let us now investigate the radial part. The general expression for a scattering event involving four different shells is expressed in (4.15)

$$\sum_{\tau_1,\tau_2,\tau_3,\tau_4} \langle R_{n_1,l_1}(r)R_{n_2,l_2}(r')| \sum_{k=0}^{min(|l_1+l_4|,|l_2+l_3|)} \frac{r_<^k}{r_>^{k+1}}|R_{n_3,l_3}(r)R_{n_4,l_4}(r')\rangle . \qquad (4.15)$$

In the case of Coulomb interaction in a single shell, $n_1 = n_2 = n_3 = n_4$ and $l_1 = l_2 = l_3 = l_4$ and hence one obtains the expression

$$F^{(k)} = \sum_{k=0}^{2l} \langle R_{n,l}(r)R_{n,l}(r')| \frac{r_<^k}{r_>^{k+1}}|R_{n,l}(r)R_{n,l}(r')\rangle . \qquad (4.16)$$

For a $3d^n$ configuration, $2l = 4$ yielding values of $k = 0, 2, 4$ and one has to evaluate three Slater integrals $F^{(k)}$ for such a configuration. For an excited state with a core hole, like the excited state of a L absorption edge with a configuration $2p^5 3d^{n+1}$, it is necessary to take into consideration the Coulomb interaction between the $2p$ and the $3d$ shells. There are two possible cases for these scattering events (see also Chap. 2):

1. Direct interaction with $n_1 = n_3$ and $n_2 = n_4$. This means that each electron scatters in its shell. The matrix elements read

$$F^{(k)} = \sum_{k=0}^{min(|2l_1,2l_2|)} \langle R_{n_1,l_1}(r)R_{n_2,l_2}(r')| \frac{r_<^k}{r_>^{k+1}}|R_{n_1,l_1}(r)R_{n_2,l_2}(r')\rangle . \qquad (4.17)$$

2. Exchange interaction with $n_1 = n_4$ and $n_2 = n_3$. This means that electrons exchange shells. The matrix elements read

$$G^{(k)} = \sum_{k=0}^{min(|2l_1,2l_2|)} \langle R_{n_1,l_1}(r)R_{n_2,l_2}(r')| \frac{r_<^k}{r_>^{k+1}} |R_{n_2,l_2}(r)R_{n_1,l_1}(r')\rangle . \tag{4.18}$$

The $F^{(k)}$ and $G^{(k)}$ are called Slater integrals. The magnitude of the multiplet splitting depends on the magnitude of the Slater integrals. For an atomic calculation (corresponding to the case of a free ion in spherical symmetry) radial integrals are calculated self-consistently using a Hartree–Fock model, with values typically of the order of a few eV (not more than tens of eV). In the point group symmetry, where the absorber is considered with its environment, these values are reduced empirically in order to take into account the effect of the chemical bond which delocalizes the electrons. This reduction factor is an adjustable parameter (typically, 60–80% for a iono-covalent bond, 100% being the ionic limit case of a free ion).

The Crystal Field Hamiltonian

Let us now extend our theoretical framework to include the effect of the CF potential on the absorbing ion. This is done by considering the N nearest neighbours as point charges ($Z_i e$) at positions $\mathbf{R_i}$. The electrostatic potential due to these point charges at position $\mathbf{r}$, $V_{CF}(\mathbf{r})$, is expressed as

$$V_{CF}(\mathbf{r}) = \sum_{i=1}^{N} \frac{Z_i e^2}{|\mathbf{r} - \mathbf{R_i}|} . \tag{4.19}$$

A multipole expansion of the potential can be used to expand (4.19), leading to the expression

$$V_{CF}(r, \theta, \phi) = e^2 \sum_{k=0}^{\infty} r^k \sum_{m=-k}^{k} C_{k,m}(\theta, \phi) Q_{k,m} , \tag{4.20}$$

with $Q_{k,m} \equiv \left[\frac{4\pi}{2k+1}\right]^{1/2} \sum_{i=1}^{N} Z_i(\frac{1}{R_i})^{k+1} Y_{k,m}^*(\theta_i, \phi_i)$. $Y_{k,m}$ is the spherical harmonic, $C_{k,m}$ is the renormalized spherical harmonic, and $*$ is the complex conjugate. Note here that one assumes that the radial extent of the $3d$ orbitals is smaller than the distance between the absorbing ion and its first neighbours (i.e., $r \ll R_i$). The CF Hamiltonian can be developed using the single-particle basis (atomic spin-orbitals as discussed before for the Coulomb interaction). Separating the radial and angular parts leads to

$$\hat{H}_{CF} = \sum_{\tau_1,\tau_2} \sum_{k,m} A_{k,m} \langle Y_{\ell_1,m_1}\sigma_1 | C_{k,m} | Y_{\ell_2,m_2}\sigma_2 \rangle c_{\tau_1}^\dagger c_{\tau_2} . \tag{4.21}$$

Here, the $A_{k,m}$ combine all the radial parts of (4.20). They are related to the usual CF parameters ($10Dq$, Ds, Dt, ...) which are usually not known precisely. The CF parameters (or $A_{k,m}$) are either fitted parameters or taken from experiments (optical absorption, electron paramagnetic resonance, ...). It is important to warn the reader against the temptation to fit the calculation with an unreasonable number of CF parameters: with adjustable parameters, one runs the risk of producing some spectra in good agreement with experiments not for a theoretically justifiable reason but rather due to some lucky cancellation between inappropriate choices of the parameters and inaccurate theoretical approximations employed in solving the model. It is therefore important that one uses some sensible limits for these parameters and critically examines the values used in the model.

Equation 4.21 is a useful expression because the matrix elements are integrals again over three spherical harmonics and are given by the Gaunt coefficients. Furthermore, the series can be truncated according to the triangular condition. For $3d$ orbitals (like in the case of an Fe ion), $\ell_1 = \ell_2 = 2$ so the maximum value of m possible is $m = 4$ with $k \le 4$. The actual form of the matrix elements depends on the symmetry of the CF potential.

We will present the example of an octahedral (O_h) cluster to illustrate the procedure of calculating the CF matrix elements. In the case of an O_h cluster, six neighbours are positioned at equal distances from the central ion as shown in Fig. 4.1. We will first calculate the multipole terms possible for this configuration according to (4.20). This can be easily evaluated and many softwares are available such as the "multipoles" Python package [8]. One finds that the multipole expansion of the octahedral potential reduces to the following terms with the coefficients listed below $C_{0,0} \rightarrow 6$,

$$C_{4,0} \rightarrow \sqrt{\frac{49}{4}}, \; C_{4,4} \rightarrow \sqrt{\frac{35}{8}}, \; C_{4,-4} \rightarrow \sqrt{\frac{35}{8}}.$$

Now we can evaluate the matrix elements of (4.21) using the Gaunt coefficients. One finds the following matrix for O_h crystal field

$$\sum_{k,m} \langle Y_{2,m_1} | C_{k,m} | Y_{2,m_2} \rangle \propto \begin{bmatrix} 1 & 0 & 0 & 0 & 5 \\ 0 & -4 & 0 & 0 & 0 \\ 0 & 0 & 6 & 0 & 0 \\ 0 & 0 & 0 & -4 & 0 \\ 5 & 0 & 0 & 0 & 1 \end{bmatrix}. \tag{4.22}$$

Diagonalizing this matrix gives the following eigenvalues $E = 6, -4, -4, 6, -4$ for the eigenvectors

$$\begin{bmatrix} \frac{1}{\sqrt{2}} \\ 0 \\ 0 \\ 0 \\ \frac{1}{\sqrt{2}} \end{bmatrix}, \begin{bmatrix} -\frac{1}{\sqrt{2}} \\ 0 \\ 0 \\ 0 \\ \frac{1}{\sqrt{2}} \end{bmatrix}, \begin{bmatrix} 0 \\ 1 \\ 0 \\ 0 \\ 0 \end{bmatrix}, \begin{bmatrix} 0 \\ 0 \\ 1 \\ 0 \\ 0 \end{bmatrix}, \begin{bmatrix} 0 \\ 0 \\ 0 \\ 1 \\ 0 \end{bmatrix}. \tag{4.23}$$

This illustrates the splitting of the $3d$ one electron orbitals in an O_h crystal field as shown in Fig. 4.1. The energy difference between the t_{2g} and e_g orbitals is referred to as $10Dq$ and its magnitude depends on the radial part $A_{k,m}$. The five degenerate orbitals split into two types of orbital:

1. Three orbitals of energies $-4Dq$ referred to as the t_{2g} orbitals and which are $\frac{1}{\sqrt{2}}(Y_{2,-2} - Y_{2,2})$, $Y_{2,1}$, and $Y_{2,-1}$.

Fig. 4.1 Splitting of the five degenerate $3d$ orbitals (left) into t_{2g} and e_g orbitals in an octahedral crystal field (right)

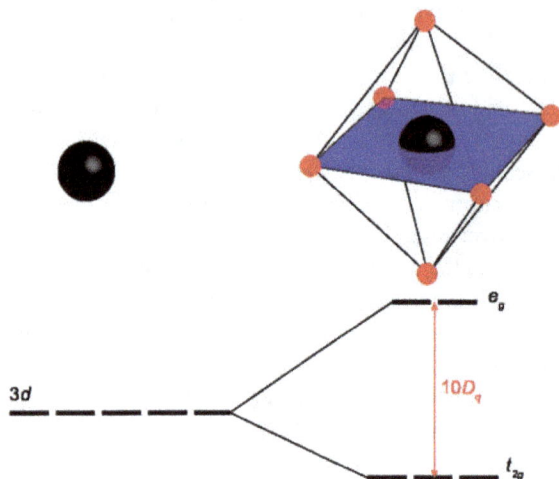

2. Two orbitals of energies $6Dq$ referred to as the e_g orbitals and which are $\frac{1}{\sqrt{2}}(Y_{2,-2} + Y_{2,2})$, and $Y_{2,0}$.

The same procedure can be used to determine the eigenstates and energies of the $3d$ (or any other) orbitals embedded in a certain symmetry.

4.1.3.3   The Many-Body Extended Picture of Electronic States

The most recently developed approaches aim at extending both DFT and LFM theories to provide a comprehensive description of many-body interactions. An example for a rather manageable improvement to DFT is to add a static Hubbard parameter U to account for on-site electronic repulsion. In addition, there are other approaches that go well beyond DFT. They can be grouped into two main types: the quantum chemistry approaches (mainly, wave function-based methods) and the Green's function methods.

Quantum chemistry approaches (DFT-CI, multi-configurational self consistent field [configuration interaction (CI)], coupled cluster, quantum Monte Carlo) are many-body extended approaches. In DFT-CI, for example, a combination of Slater determinants is used to describe the wave function of the system. The number of configurations that can be considered is limited by computing power and one has to decide which configurations to include. These approaches can only be applied for small clusters and molecules due to computational demand. Green's functions based methods, such as GW and dynamical mean-field theory (DMFT) provide an alternative approach to calculate the electronic structure of strongly correlated materials. In a GW method, a screened Coulomb interaction (W) is calculated following a DFT calculation of the charge density. A nonlocal energy dependent self-energy operator is required. Furthermore, DMFT can be utilized to map the full lattice problem onto a single site quantum impurity problem, using a local screened Coulomb interaction (U). In both GW and DMFT approaches, correlations (many-body effects) are significantly better described than in standard DFT.

4.1.3.4 Which Approach Works Best for Core Level Spectroscopy Calculations?

Unfortunately this is a very complex question: it depends on the chemical system (ionic, covalent, strongly correlated, ...), the edge (K-, L-edges, ...) and the type of spectroscopy (absorption, photoemission, ...). Consequently, in the following we try to provide some guiding considerations.

A good starting criterion for the choice of the method of calculation is the localization of the final state wave function. When electrons are excited into highly delocalized orbitals, they interact much less with other electrons meaning that the intra-atomic interactions are expected to be small. In this case single-particle approaches often give satisfactory results as the multi-electronic effects are small. This may be the case for the K edge of $3d$ elements and the $L_{2,3}$ edges of heavy elements (e.g., rare earths, $5d$ transition metals), as well as the K edges of ligands (such as C, N, O, S).

On the other hand, when electrons are excited into localized orbitals, they will strongly interact with each other and the core hole. In this case, the multi-electronic effects become significant, and description considering only the absorber atom may be more successful such as LFM theory. This is typically the case for $L_{2,3}$ edges of $3d$ elements and $M_{4,5}$ edges of $4f$ elements.

In intermediate cases where both intra- and inter-atomic effects are relevant, like the K pre-edge of $3d$ elements, single-particle and many-body approaches may work or fail. It may also happen that different energy ranges of a spectrum are best described by different theoretical approaches. This is often the case when a weak pre-edge feature is observed before the strong main edge. The pre-edges often arise from excitations into localized orbitals while the main edge has a more delocalized character. Whether the final state is localized or not may be derived from the shape of the absorption edge. When the edge is dominated by a step function, the final state can be assumed to be delocalized. If it exhibits distinct peaks that decrease sharply, this may be due (but not necessarily) to a localized final state.

4.1.4 Codes for Ligand-Field Multiplet Calculations

Regarding the practical aspects of calculating core level spectra, one faces yet another level of complexity as the plethora of theoretical methods mentioned above are implemented in a comparably large number of computational packages. Instead of trying to provide an overview of all the available computational tools, we will focus on introducing Quanty, one of the currently available software packages that can be used to perform ligand-field multiplet calculations, and Crispy a graphical user interface that uses Quanty as a computational engine.

Quanty [9–12] is developed by Maurits Haverkort and his collaborators at the Institute of Theoretical Physics at Heidelberg University. It is a computational library that can be used to write quantum mechanical programs in the second quantization formalism. While the library can be used to describe a wide range of problems, it is specifically aimed at calculating different spectroscopies, including core level spectroscopy. Briefly, the user starts by constructing the Hamiltonian for the system of interest, diagonalize it, selects several eigenstates, and then calculates the spectrum corresponding to these eigenstates. In Quanty the Hamiltonian is expressed in a basis of one particle modes, which can be both fermionic and bosonic. The fermionic modes are usually spin-orbitals. In semi-empirical multiplet calculations, the interactions between the spin-orbitals are parametrized using values calculated for isolated atoms, which are afterwards scaled to account for the effect of the surrounding atoms. Alternatively, the parameters can be calculated directly by using for example DFT-based methods [9]. After the diagonalization of the Hamiltonian and the selection of the lowest eigenstates, using, for example, Boltzmann statistics, Quanty can be used to calculate the spectrum. As mentioned previously, this is done using a Green's function approach, thereby avoiding the sum-over-states calculation, which can lead to an important reduction in computational time.

The core of the library is written using the C/C++ programming language for maximum efficiency. The users do not interact directly with this part of the code, but rather with the Lua-based layer that wraps it. To run calculations the users are required to write small programs using the functions defined in Quanty. Doing this in a scripting language such as Lua has the advantage of providing an ideal environment for experimentation, circumventing the limitations of compiled programming languages such as C or C++. While this is indeed very helpful, it is not uncommon for such programs to reach more than a few tens of lines of code, which in itself can be intimidating for the majority of new users. Also, as it is the case for many scientific libraries, because of the flexibility given to the users when writing these programs, it is impossible to check for all the things that might be incorrect. This leads to errors that are difficult to trace even for experienced users.

To help users to more easily perform Quanty calculations, one of us (M. Retegan) has developed a friendly user interface that exposes the library's capabilities for a large part of core level spectroscopies. Crispy [13] was developed using the Python programming language and relies on additional packages from the Python ecosystem. The main window of the application is shown in Fig. 4.2. Using Crispy, the users can quickly adjust the parameters of the calculation, run it, and plot the resulting spectrum without the need of writing any programs. The approach has many advantages for novice users, but even experienced ones can use Crispy to generate a starting program that will become the basis of their calculation. Crispy is a free and open-source program that can be installed on any operating system that has an up-to-date Python distribution. For Windows® and macOS® the program comes in easy to install packages that can be downloaded from the official website, http://www.esrf.eu/computing/scientific/crispy.

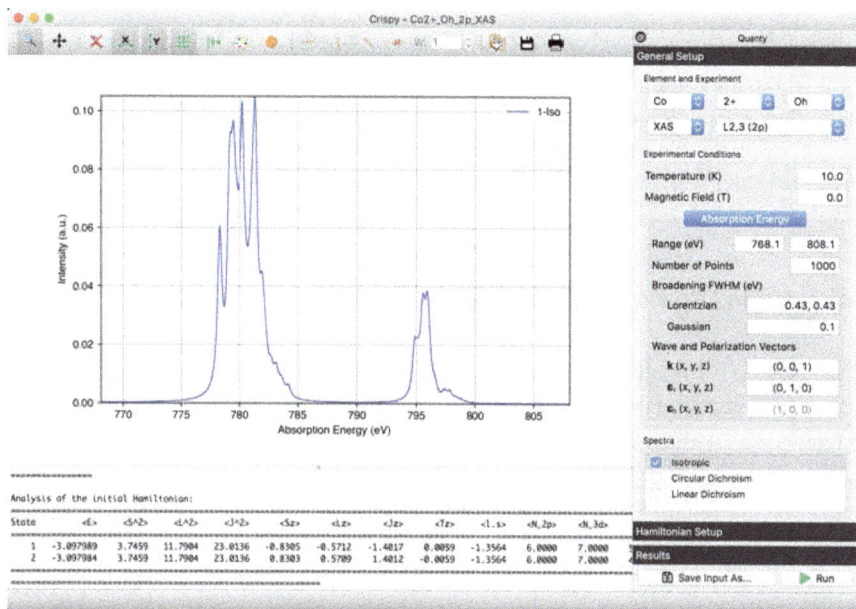

Fig. 4.2 Crispy's main window showing a calculated XAS $L_{2,3}$ spectrum for a Co^{2+} ion in octahedral symmetry

4.2 Spherical Tensor Expansion of the XAS Cross Section

A spherical tensor is a set of components that transform into each others under arbitrary rotations. Another way to state this is to say that the components of a spherical tensor generate a vector space which is invariant under rotation. A spherical tensor is irreducible if this vector space cannot be written as the sum of two invariant (non-zero) subspaces. For an irreducible tensor of rank j, the dimension of the corresponding vector space is $2j + 1$. For example, spherical harmonics $Y_{l,m}$ are the spherical tensor components of a spherical tensor of rank l. Note that while a 3×3 matrix is an irreducible Cartesian tensor, it is a reducible spherical tensor which is the sum of $j = 0$, $j = 1$, and $j = 2$ irreducible spherical tensors. It is evident that such an expansion would provide us with deeper insights by identifying groups of spectra that obey certain symmetry transformation rules which one could easily relate back to the system symmetry [14].

Spherical tensor analysis has been used with great success for the X-ray photoelectron of localized magnetic systems [15–19] and in XAS [14, 20–22], including XNLD [23]. The underlying idea is to determine a finite set of fundamental spectra in terms of which all possible experimental spectra can be expressed. More precisely, the XAS spectrum obtained for a given polarization vector (ϵ) and wave vector (k) of the incident beam is written as a sum of terms which are fundamental spectra [21, 22] (depending only on the sample properties) multiplied by an angular coefficient depending only on the experimental conditions (k, ϵ). Such a geometric and fully decoupled expression is useful: (i) to disentangle the properties of the sample from those of the measurement; (ii) to determine specific experimental arrangements aiming at the observation of specific sample properties; (iii) to provide the most convenient starting point to investigate the reduction of the number of fundamental spectra due to crystal symmetries.

4.2.1   The Case of Electric Dipole Transitions

The first step is to build rank one spherical tensors from the vectors appearing in the transition operator. The polarization vector $\epsilon = [\epsilon_x, \epsilon_y, \epsilon_z]$ can be written as a spherical tensor ϵ^1 with components $\epsilon^1_{-1} = \frac{\epsilon_x - i\epsilon_y}{\sqrt{2}}$, $\epsilon^1_1 = -\frac{\epsilon_x + i\epsilon_y}{\sqrt{2}}$, and $\epsilon^1_0 = \epsilon_z$. Similarly, the position spherical tensor, r^1, can be constructed. In the following we shall use the following notation for the coupling of spherical tensors P^a and Q^b of ranks a and b into a spherical tensor of rank c

$$\{P^a \otimes Q^b\}^c_\gamma = \sum_{\alpha=-a}^{a} \sum_{\alpha=-b}^{b} (a\alpha b\beta | c\gamma) P^a_\alpha Q^b_\beta, \tag{4.24}$$

with $(a\alpha b\beta | c\gamma)$ being the Clebsch–Gordan coefficients. Therefore,

$$P^a \cdot Q^a = \sum_{\alpha=-a}^{a} (-1)^\alpha P^a_{-\alpha} Q^a_\alpha = (-1)^a \sqrt{2a + 1}\{P^a \otimes Q^a\}^0. \tag{4.25}$$

One has now to compute the scalar product of both tensors which is given by (4.25). The dipole transition operator can be written as in (4.26) taking into consideration that r is real while ϵ is in general complex

$$\begin{aligned}
T &= -\sqrt{3}\{\epsilon^1 \otimes r^1\}^0 \\
T^\dagger &= -\sqrt{3}\{\epsilon^{1*} \otimes r^1\}^0 .
\end{aligned} \tag{4.26}$$

We can recouple the cross section such that polarization tensors are coupled to each other and position tensors are coupled to each other. This means that the expression will have a part that depends only on the experimental geometry (polarization vector) and a part that depends only on the sample properties. This recoupling can be done using the identity

$$\{P^g \otimes Q^g\}^0 \cdot \{R^d \otimes S^d\}^0 = \sum_a (-1)^a \frac{\{P^g \otimes R^d\}^a \cdot \{Q^g \otimes S^d\}^a}{\sqrt{(2g + 1)(2d + 1)}} . \tag{4.27}$$

Here a is constrained to $|g - d| \le a \le g + d$. Hence, for the dipole transition, $g = d = 1$ and $0 \le a \le 2$. The recoupled XAS cross section is finally expressed as follows:

$$\sigma_\omega = -4\pi\alpha\hbar\omega\text{Im}\left[\sum_{a=0}^{2}(-1)^a\{\epsilon^{1*}\otimes\epsilon^1\}^a \cdot \{\langle I|r^1 G^+ r^1|I\rangle\}^a\right]. \quad (4.28)$$

Quanty can calculate the energy dependent tensors $R^{(a)} = \{\langle I|r^1 G^+ r^1|I\rangle\}^a$ that depend only on the properties of the sample. We will refer to these elements as the fundamental spectra. Note that these fundamental spectra are sometimes referred to as $\sigma^{(a)}$. This could be confused with the total cross section σ_ω so we shall not use this notation here. The experimental geometry tensor is $E^a = \{\epsilon^{1*}\otimes\epsilon^1\}^a$.

4.2.1.1 Term $a = 0$

The first term can be found by substituting $a = 0$ in (4.28). This is the zero rank of the tensor, given in (4.29).

$$\sigma(0,0) = 4\pi\alpha\hbar\omega \times \text{Im}\left[\frac{1}{3}\left(\langle I|rC_{1,0}^* G^+ rC_{1,0}|I\rangle + \langle I|rC_{1,-1}^* G^+ rC_{1,-1}|I\rangle\right.\right.$$
$$\left.\left.+\langle I|rC_{1,1}^* G^+ rC_{1,1}|I\rangle\right)\right]. \quad (4.29)$$

Here $C_{\ell m} = C_m^\ell = \sqrt{\frac{4\pi}{2\ell+1}}Y_\ell^m$. The term $\sigma(0,0)$ is independent of the incident polarization vector and as such is rotation invariant. It gives the isotropic contribution of the XAS cross section.

4.2.1.2 Term $a = 1$

The term $a = 1$ consists of three components, namely, $\sigma(1,0), \sigma(1,1)$, and $\sigma(1,-1)$:

$$\sigma(1,0) = -4\pi\alpha\hbar\omega \times \text{Im}\left[\frac{1}{2}\left(i\epsilon_x^*\epsilon_y - i\epsilon_x\epsilon_y^*\right)\right.$$
$$\left.\times\left(\langle I|rC_{1,1}^* G^+ rC_{1,1}|I\rangle - \langle I|rC_{1,-1}^* G^+ rC_{1,-1}|I\rangle\right)\right], \quad (4.30)$$

$$\sigma(1,1) = -4\pi\alpha\hbar\omega \times \text{Im}\left[\frac{-1}{2\sqrt{2}}\left(\epsilon_x^*\epsilon_z - \epsilon_x\epsilon_z^* + i\epsilon_y\epsilon_z^* - i\epsilon_y^*\epsilon_z\right)\right. \tag{4.31}$$

$$\left.\left(\langle I|rC_{1,0}^*G^+rC_{1,1}|I\rangle + \langle I|rC_{1,-1}^*G^+rC_{1,0}|I\rangle\right)\right], \tag{4.32}$$

$$\sigma(1,-1) = -4\pi\alpha\hbar\omega \times \text{Im}\left[\frac{1}{2\sqrt{2}}\left(\epsilon_x^*\epsilon_z - \epsilon_x\epsilon_z^* + i\epsilon_y^*\epsilon_z - i\epsilon_y\epsilon_z^*\right)\right.$$
$$\left.\times\left(\langle I|rC_{1,0}^*G^+rC_{1,-1}|I\rangle + \langle I|rC_{1,1}^*G^+rC_{1,0}|I\rangle\right)\right]. \tag{4.33}$$

One notices from (4.30), (4.32) and (4.33) that the spectra of $a = 1$ are not active if any of these two cases are satisfied:

1. If linearly polarized light is used for the measurement.
2. If all off-diagonal elements are zero and the diagonal elements are equal.

Another conclusion that can be drawn from the recoupling is the necessity to perform XAS measurements using both linearly and circularly polarized light to probe these fundamental spectra.

4.2.1.3 Term $a = 2$

The term $a = 2$ consists of five components, namely, $\sigma(2,0)$, $\sigma(2,1)$, $\sigma(2,-1)$, $\sigma(2,2)$, and $\sigma(2,-2)$. These are given below as follows:

$$\sigma(2,0) = -4\pi\alpha\hbar\omega \times \text{Im}\left[\frac{1}{6}\left(2|\epsilon_z|^2 - |\epsilon_x|^2 - |\epsilon_y|^2\right)\left(2\langle I|rC_{1,0}^*G^+rC_{1,0}|I\rangle\right.\right.$$
$$\left.\left. -\langle I|rC_{1,-1}^*G^+rC_{1,-1}|I\rangle - \langle I|rC_{1,1}^*G^+rC_{1,1}|I\rangle\right)\right], \tag{4.34}$$

$$\sigma(2,1) = -4\pi\alpha\hbar\omega \times \text{Im}\left[\frac{1}{2\sqrt{2}}\left(\epsilon_x\epsilon_z^* + \epsilon_x^*\epsilon_z - i\epsilon_y\epsilon_z^* - i\epsilon_y^*\epsilon_z\right)\right.$$
$$\left.\times\left(\langle I|rC_{1,-1}^*G^+rC_{1,0}|I\rangle - \langle I|rC_{1,0}^*G^+rC_{1,1}|I\rangle\right)\right], \tag{4.35}$$

$$\sigma(2,-1) = -4\pi\alpha\hbar\omega \times \text{Im}\left[\frac{1}{2\sqrt{2}}\left(\epsilon_x\epsilon_z^* + \epsilon_x^*\epsilon_z + i\epsilon_y^*\epsilon_z + i\epsilon_y\epsilon_z^*\right)\right.$$
$$\left.\times\left(\langle I|rC_{1,0}^*G^+rC_{1,-1}|I\rangle - \langle I|rC_{1,1}^*G^+rC_{1,0}|I\rangle\right)\right], \tag{4.36}$$

$$\sigma(2, 2) = -4\pi\alpha\hbar\omega \times \text{Im}\left[\frac{-1}{2}\left((\epsilon_x - i\epsilon_y)(\epsilon_x^* - i\epsilon_y^*)\right)\right.$$
$$\left. \times\left(\langle I|rC_{1,-1}^* G^+ rC_{1,1}|I\rangle\right)\right], \tag{4.37}$$

$$\sigma(2, -2) = -4\pi\alpha\hbar\omega \times \text{Im}\left[\frac{-1}{2}\left((\epsilon_x + i\epsilon_y)(\epsilon_x^* + i\epsilon_y^*)\right)\right.$$
$$\left. \times\left(\langle I|rC_{1,1}^* G^+ rC_{1,-1}|I\rangle\right)\right]. \tag{4.38}$$

It can be noted from (4.34), (4.35), (4.36), (4.37), and (4.38) that the $a = 2$ spectra are active for linearly polarized light and hence these spectra are responsible for the angular dependence observed with linear light. On the contrary, no difference can be observed between right and left circularly polarized light. Another feature of these terms is that they are not active if the following two conditions are satisfied:

1. The diagonal matrix elements are equal.
2. The off-diagonal matrix elements are zero.

4.2.1.4 General Dipole Expression

The general dipole expression is given in (4.39). From this equation, the dipole XAS cross-section for an arbitrary polarization (ϵ) can be constructed from the nine fundamental spectra derived above.

$$\sigma_\omega^{Dipole}(\epsilon) = -4\pi\alpha\hbar\omega \times \text{Im}\left[\frac{1}{3}R(0, 0) + \frac{1}{2}\left(i\epsilon_x^*\epsilon_y - i\epsilon_x\epsilon_y^*\right)R(1, 0)\right.$$
$$-\frac{1}{2\sqrt{2}}\left(\epsilon_x^*\epsilon_z - \epsilon_x\epsilon_z^* + i\epsilon_y\epsilon_z^* - i\epsilon_y^*\epsilon_z\right)R(1, 1)$$
$$+\frac{1}{2\sqrt{2}}\left(\epsilon_x^*\epsilon_z - \epsilon_x\epsilon_z^* + i\epsilon_y^*\epsilon_z - i\epsilon_y\epsilon_z^*\right)R(1, -1)$$
$$+\frac{1}{6}\left(2|\epsilon_z|^2 - |\epsilon_x|^2 - |\epsilon_y|^2\right)R(2, 0)$$
$$+\frac{1}{2\sqrt{2}}\left(\epsilon_x\epsilon_z^* + \epsilon_x^*\epsilon_z - i\epsilon_y\epsilon_z^* - i\epsilon_y^*\epsilon_z\right)R(2, 1)$$
$$+\frac{1}{2\sqrt{2}}\left(\epsilon_x\epsilon_z^* + \epsilon_x^*\epsilon_z + i\epsilon_y^*\epsilon_z + i\epsilon_y\epsilon_z^*\right)R(2, -1)$$
$$-\frac{1}{2}\left((\epsilon_x - i\epsilon_y)(\epsilon_x^* - i\epsilon_y^*)\right)R(2, 2)$$
$$\left. -\frac{1}{2}\left((\epsilon_x + i\epsilon_y)(\epsilon_x^* + i\epsilon_y^*)\right)R(2, -2)\right], \tag{4.39}$$

where the R are the fundamental spectra and are defined as

$$R(0,0) = \langle I|rC_{1,0}^*G^+rC_{1,0}|I\rangle + \langle I|rC_{1,-1}^*G^+rC_{1,-1}|I\rangle$$
$$+\langle I|rC_{1,1}^{1*}G^+rC_{1,1}|I\rangle, \tag{4.40}$$

$$R(1,0) = \langle I|rC_{1,1}^*G^+rC_{1,1}|I\rangle - \langle I|rC_{1,-1}^*G^+rC_{1,-1}|I\rangle, \tag{4.41}$$

$$R(1,1) = \langle I|rC_{1,0}^*G^+rC_{1,1}|I\rangle + \langle I|rC_{1,-1}^*G^+rC_{1,0}|I\rangle, \tag{4.42}$$

$$R(1,-1) = \langle I|rC_{1,0}^*G^+rC_{1,-1}|I\rangle + \langle I|rC_{1,1}^*G^+rC_{1,0}|I\rangle, \tag{4.43}$$

$$R(2,0) = 2\langle I|rC_{1,0}^*G^+rC_{1,0}|I\rangle - \langle I|rC_{1,-1}^*G^+rC_{1,-1}|I\rangle$$
$$-\langle I|rC_{1,1}^*G^+rC_{1,1}|I\rangle, \tag{4.44}$$

$$R(2,1) = \langle I|rC_{1,-1}^*G^+rC_{1,0}|I\rangle - \langle I|rC_{1,0}^*G^+rC_{1,1}|I\rangle, \tag{4.45}$$

$$R(2,-1) = \langle I|rC_{1,0}^*G^+rC_{1,-1}|I\rangle - \langle I|rC_{1,1}^*G^+rC_{1,0}|I\rangle, \tag{4.46}$$

$$R(2,2) = \langle I|rC_{1,-1}^*G^+rC_{1,1}|I\rangle, \tag{4.47}$$

$$R(2,-2) = \langle I|rC_{1,1}^*G^+rC_{1,-1}|I\rangle. \tag{4.48}$$

4.2.1.5   Case Study of a d^9 ion

Octahedral Crystal Field

Equation (4.39) can be simplified when the symmetry of the absorbing system is taken into account. As a demonstration, we shall study a d^9 ion in octahedral (O_h) symmetry. Figure 4.3 (top) shows the matrix elements for such a system. These matrix elements are the direct output of Quanty and will be referred to as the conductivity tensor. One finds that all the off-diagonal matrix elements are equal to zero and all the diagonal matrix elements are equal. This leaves only the $R(0,0)$ term of (4.39) not equal to zero. Hence one can conclude that the cross section of a dipole transition is isotropic in an O_h system.

Tetragonal Crystal Field

Let us now consider a tetragonal distortion such that the octahedron is compressed along the z-axis. The ground state in this case has a hole in the d_{z^2} orbital (neglecting spin-orbit coupling) and the z-axis is now different from the x- and y-axes. The conductivity tensor for such a system is shown in Fig. 4.3 (bottom). As could be intuitively expected, the middle panel corresponding to $C_0^1G^+C_0^1$ is different from the other two diagonal elements. This implies that the following terms come into play (see Fig. 4.4):

- $R(0,0)$ which gives the isotropic cross section.
- $R(2,0)$ which has a polarization dependence of the form $\frac{1}{6}\left(2|\epsilon_z|^2 - |\epsilon_x|^2 - |\epsilon_y|^2\right)$.

It is interesting in this case to investigate what types of dichroism effect could be observed. Consider rotating the incident linear polarization vector in the $x - y$-plane.

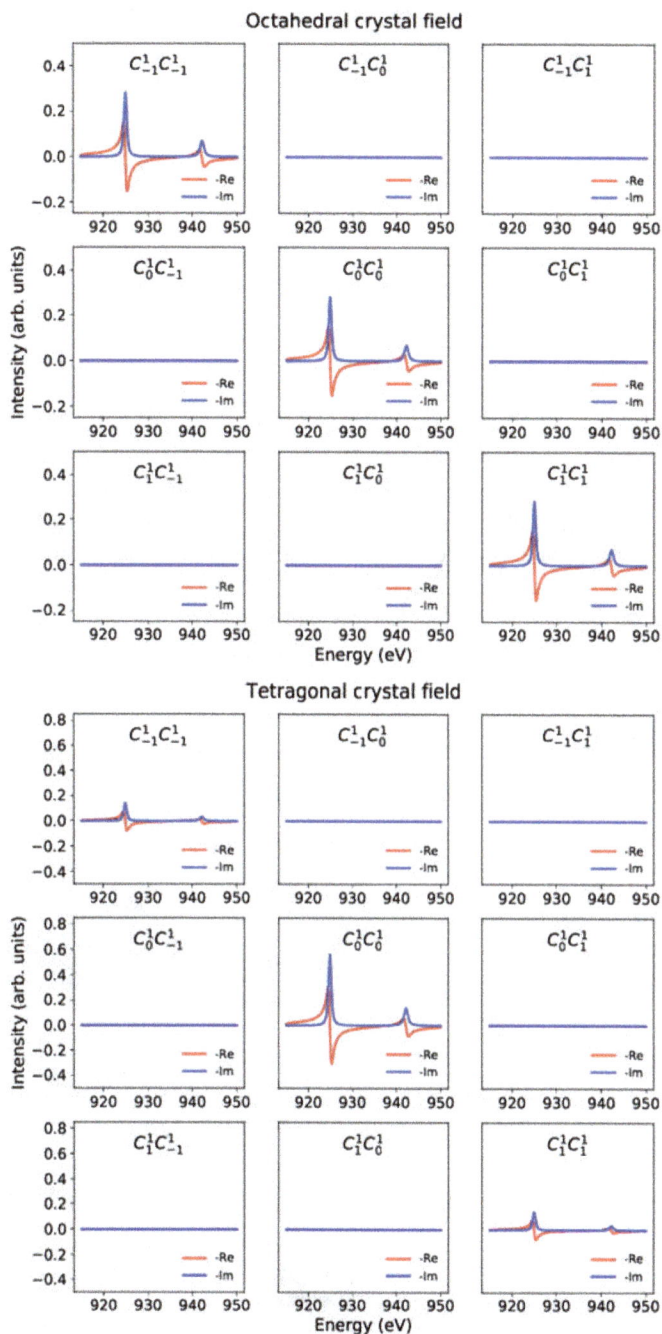

Fig. 4.3 Conductivity tensor calculated for a d^9 ion in an octahedral crystal field (top) and in a tetragonal crystal field (bottom). Calculations are done with a crystal field parameter $10D_q = 1.1\,\text{eV}$ and with $D_s = -0.2\,\text{eV}$ in tetragonal symmetry. Re and Im are the real and imaginary parts of the tensor

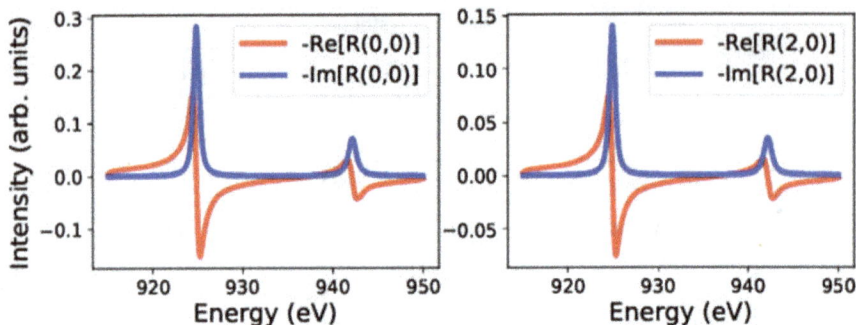

Fig. 4.4 Fundamental spectra $R(0, 0)$ and $R(2, 0)$ for a dipole transition for a d^9 ion in a tetragonal crystal field ($10D_q = 1.1$ eV and $D_s = -0.2$ eV)

In this case $\epsilon = [\cos(\theta), \sin(\theta), 0]$ where θ is the rotation angle defined from the x-axis. The expression of the polarization dependence for the term $R(2, 0)$ reveals that no angular dependence is to be expected in this case. The system is effectively O_h in the $x - y$-plane and one would expect no angular dependence as discussed in the previous example. This XAS cross section for this rotation is shown in Fig. 4.5a. On the contrary, rotating the polarization vector in the $x - z$ ($\epsilon = [\cos(\theta), 0, \sin(\theta)]$) or $y - z$ ($\epsilon = [0, \cos(\theta), \sin(\theta)]$) planes should yield an angular dependence as the polarization vector probes the distortion, which is indeed observed as shown in Fig. 4.5b and c. The dependence of the XAS cross-section on the direction of the

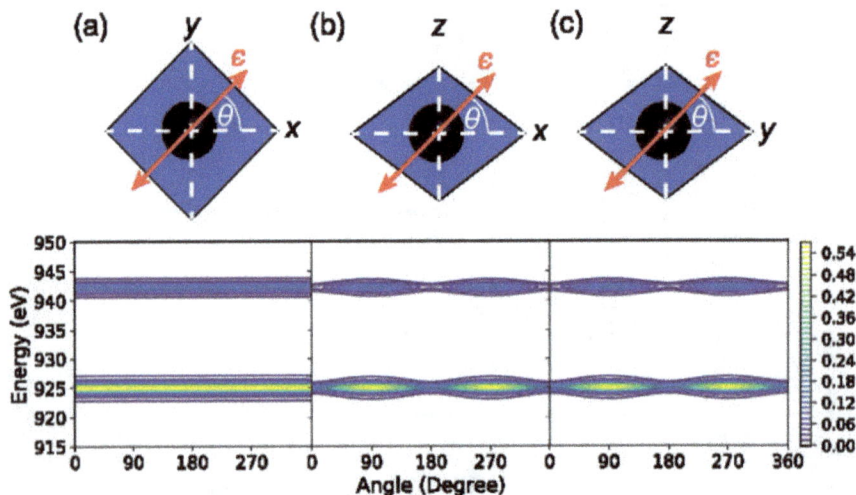

Fig. 4.5 Angular dependence of the $L_{2,3}$ XAS of a $3d^9$ ion in a tetragonal crystal field. Calculations are done by rotating the polarization vector in the $x - y$-plane [panel (**a**)], $x - z$-plane [panel (**b**)], and $y - z$-plane [panel (**c**)] as illustrated in the top panel

linearly polarized light is an effect referred to as linear dichroism as discussed previously.

Octahedral Crystal Field with an Exchange Field ∥ z

Another interesting system to investigate is a magnetic $3d^9$ ion where the crystal field is O_h with an exchange field aligned along the z-axis. Hence, the z-axis is inequivalent to the x- and y-axes due to the exchange field. The conductivity tensor of such a system is shown in Fig. 4.6. The exchange field is aligned along a high symmetry direction in this example which preserves the C_4 rotation symmetry of the system and consequently preserves the symmetry of the conductivity tensor. All off-diagonal elements are zero. Note that the off-diagonal elements are zero because we chose to calculate the tensor using the symmetry adapted transition operators. Three fundamental spectra come into play and are plotted in Fig. 4.7:

- $R(0, 0)$ which gives the isotropic cross section,
- $R(1, 0)$ which has a polarization dependence of the form $\frac{1}{2}\left(i\epsilon_x^*\epsilon_y - i\epsilon_x\epsilon_y^*\right)$,
- $R(2, 0)$ which has a polarization dependence of the form $\frac{1}{6}\left(2|\epsilon_z|^2 - |\epsilon_x|^2 - |\epsilon_y|^2\right)$.

Nearly no angular dependence can be observed by rotating the incident linear polarization vector in the $x - y$-, $x - z$-, and $y - z$-planes (see Fig. 4.8a, b, and c). This is consistent with the fact that the fundamental spectrum $R(2, 0)$ responsible for the angular dependence is nearly zero [$R(2, 0)$ is about two orders of magnitude smaller than the other two fundamental spectra in this system]. The difference in the absorption cross section of linear polarized light in a magnetic system is an effect referred to as XMLD [24]. The magnitude of the XMLD effect for this system can be seen in Fig. 4.9. Note that the magnitude of the XMLD effect in Fe_3O_4 is ~1% of the XAS signal, which could be reliably measured on existing beamlines [25].

A strong dichroism is observed when circularly polarized light is used as in the case for Fig. 4.10a. Here the incident polarization vector is either left or right polarized about the z-axis leading to a difference in the absorption. This is an effect referred to as X-ray magnetic circular dichroism (XMCD) [26]. It can be seen from the expression of the polarization part of the cross-section that if the incident wave vector is aligned perpendicular to the exchange field, for example, for $\epsilon = [0, -\frac{i}{\sqrt{2}}, \frac{1}{\sqrt{2}}]$, and $\epsilon = [0, -\frac{i}{\sqrt{2}}, -\frac{1}{\sqrt{2}}]$, no XMCD effect is observed. This is shown in Fig. 4.10b and c. This dichroism can be used to quantify the ground state spin and orbital moments of the system as given by the sum rules [27].

Octahedral Crystal Field with an Exchange Field ∥ [210]

As a last example, we consider a system in C_1 symmetry. Consider aligning the exchange field along a low symmetry direction, e.g., [210]. The exchange field now completely breaks the symmetry of the system and the conductivity tensor has off-diagonal elements (bottom of Fig. 4.6). Contrary to the previous case (where the exchange field was aligned to the z-axis), it is now not possible to find a rotated

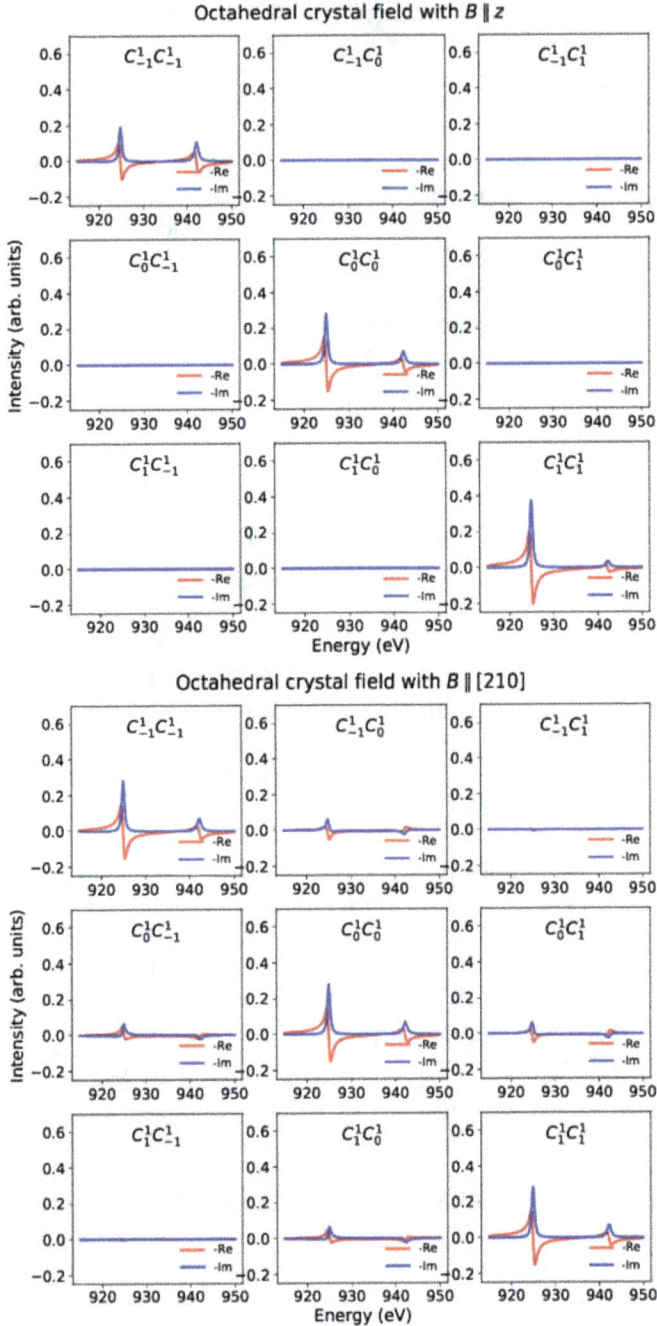

Fig. 4.6 Conductivity tensor calculated for a d^9 ion in an octahedral crystal field ($10D_q = 1.1$ eV) with an exchange field ($B = 0.05$ eV) along the z-axis (top) and along the [210] direction (bottom). The calculations are performed using a real spherical harmonics basis set

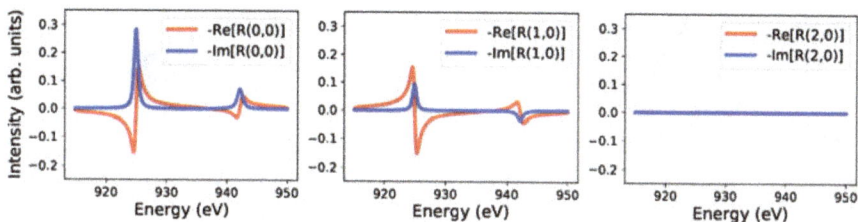

Fig. 4.7 Fundamental spectra $R(0, 0)$, $R(1, 0)$, and $R(2, 0)$ for a dipole transition for a d^9 ion in an octahedral crystal field ($10D_q = 1.1$ eV) with an exchange field ($B = 0.05$ eV) aligned along the z-axis of the cluster

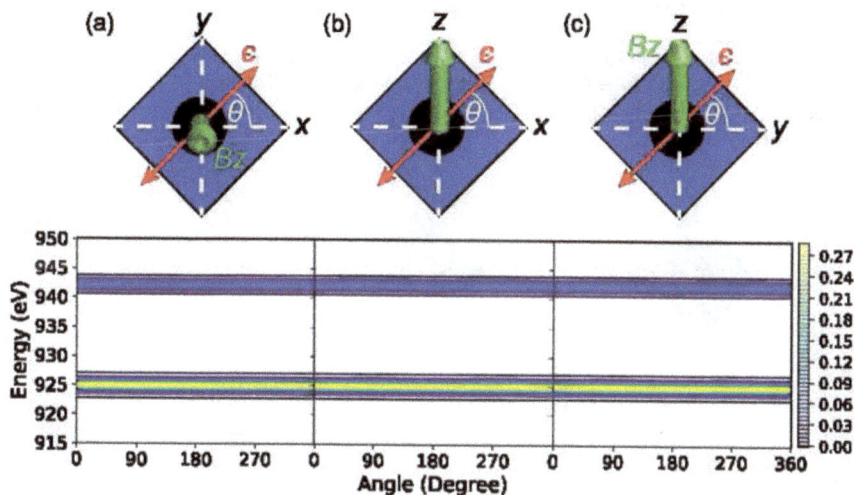

Fig. 4.8 Angular behaviour of the $L_{2,3}$ XAS of a $3d^9$ ion in an octahedral crystal field ($10D_q = 1.1$ eV) with the exchange field ($B_z = 0.05$ eV) aligned along the z-axis. The calculations are done by rotating the polarization vector in the **a** $x - y$-plane, **b** $x - z$-plane and **c** $y - z$-plane as depicted in the top panel

basis set that diagonalizes the conductivity tensor for all excited states (i.e., the basis set becomes energy dependent). It remains possible to diagonalize the conductivity tensor for a given excited state. It is important to realize that when the exchange field is aligned along a low symmetry direction, off-diagonal elements become important and more dichroic effects come into play according to (4.39) and (4.48).

4.2.2 The Case of Electric Quadrupole Transitions

For electric quadrupole transitions, we will follow the same procedure as the one used for electric dipole. The transition operator is now (up to a factor of $i/2$) $T =$

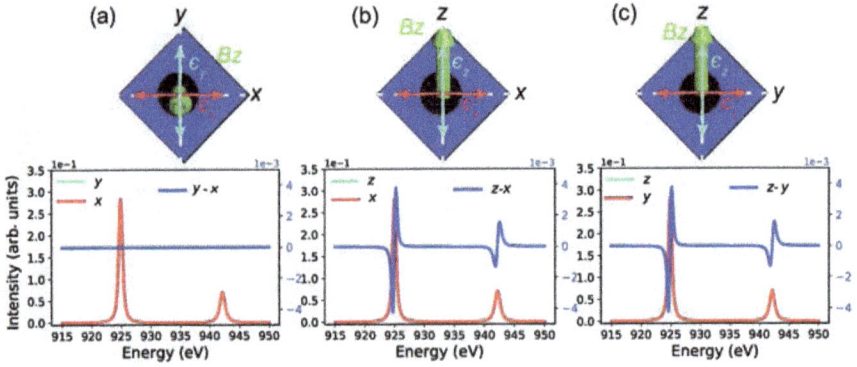

Fig. 4.9 X-ray magnetic linear dichroism of a $3d^9$ ion in an octahedral crystal field ($10D_q = 1.1$ eV) with the exchange field ($B_z = 0.05$ eV) aligned along the z-axis. The dichroism is computed by subtracting the XAS: **a** with $\epsilon \parallel x$ from that with $\epsilon \parallel y$, **b** with $\epsilon \parallel x$ from that with $\epsilon \parallel z$ and **c** with $\epsilon \parallel y$ from that with $\epsilon \parallel z$

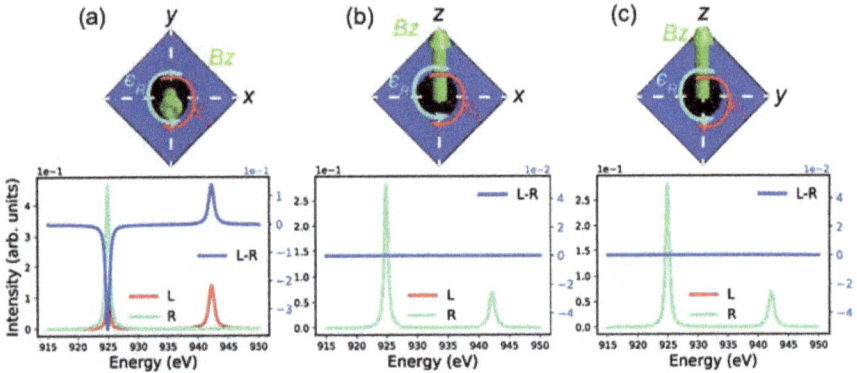

Fig. 4.10 X-ray magnetic circular dichroism of a $3d^9$ ion in an octahedral surrounding with the exchange field (B_z) aligned along the z-axis. The dichroism is computed by subtracting the XAS signal calculated with right circularly polarized light from that with left polarized light. **a** The incident wave vector is aligned parallel to the z-axis. **b** The incident wave vector is aligned parallel to the y-axis. **c** The incident wave vector is aligned parallel to the x-axis

$(\epsilon \cdot r)(k.r)$. It can be seen from the expression of the transition operator that the cross section will depend on the orientation of the polarization vector (ϵ) and of the wave vector (k) with respect to the absorbing system. Two recoupling steps are required in this case. First, the transition operator can be rewritten into a combination of scalar products of two tensors: one tensor that depends only on ϵ and k coupled together, and one tensor that depends only on the absorber r. This recoupled transition operator is expressed as follows:

$$\hat{T} = \sum_{b=0}^{2}(-1)^b\{\epsilon^1 \otimes k^1\}^b.\{r^1 \otimes r^1\}^b \ ,$$

$$(4.49)$$

$$\hat{T}^\dagger = \sum_{c=0}^{2}(-1)^c\{\epsilon^{1*} \otimes k^1\}^c.\{r^1 \otimes r^1\}^c \ .$$

The next step is to recouple the two transition amplitudes of the absorption cross section. This gives the expression

$$\sigma_\omega = \pi^2\alpha\hbar\omega \times \text{Im}\left[\sum_{a=0}^{4}\sum_{b=0}^{2}\sum_{c=0}^{2} (-1)^a(-1)^b(-1)^c\{\{\epsilon^{*1} \otimes k^1\}^c \otimes \{\epsilon^1 \otimes k^1\}^b\}^a \right.$$

$$\left. \times \{\langle I|\{r^1 \otimes r^1\}^c G^+\{r^1 \otimes r^1\}^b|I\rangle\}^a\right] \ . \quad (4.50)$$

Before attempting to write out the recoupled absorption cross section in (4.50), it is useful to simplify the expression of the transition operator first. This in turn will simplify the expression of the absorption cross section. The transition operator is a rank two tensor according to (4.49) with $b = 0, 1, 2$. We shall write out the three b terms:

- **Term $b = 0$**

$$(-1)^0\{\epsilon^1 \otimes k^1\}^0.\{r^1 \otimes r^1\}^0 = \left(-\frac{1}{\sqrt{3}}\epsilon_0^1 k_0^1 + \frac{1}{\sqrt{3}}\epsilon_1^1 k_{-1}^1 + \frac{1}{\sqrt{3}}\epsilon_{-1}^1 k_1^1\right)$$

$$\times \left(-\frac{1}{\sqrt{3}}r_0^1 r_0^1 + \frac{1}{\sqrt{3}}r_1^1 r_{-1}^1 + \frac{1}{\sqrt{3}}r_{-1}^1 r_1^1\right) \ (4.51)$$

The first part of the expression can be rewritten as $\frac{1}{\sqrt{3}}\left(-\epsilon_z k_z - \epsilon_x k_x - \epsilon_y k_y\right)$. This is equal to zero because the polarization vector is orthogonal to the wave vector. This means that the term $b = 0$ is zero.

- **Term $b = 1$**
  This term consists of three components according to

$$(-1)^1\{\epsilon^1 \otimes k^1\}^1.\{r^1 \otimes r^1\}^1 = -\frac{i}{\sqrt{2}}\frac{i}{\sqrt{2}}(\epsilon \times k) \cdot (r \times r) \ . \quad (4.52)$$

The second part of the expression is equal to zero because it is a cross product of the same vector. This means that the term $b = 1$ is also zero.

- **Term $b = 2$**
  This term consists of five components. These five components can be simplified applying the orthogonality between ϵ and k. In addition the r tensor can be expressed in terms of spherical harmonics of $l = 2$ according to the relation

$\{r^1 \otimes r^1\}_m^2 = r_m^2 = \sqrt{\frac{8\pi}{15}} Y_{2,m}(r) = \sqrt{\frac{8\pi}{15}} r^2 Y_{2,m}(\theta, \phi)$. One obtains the following five components after simplification

$$\{\epsilon^1 \otimes k^1\}_0^2 \{r^1 \otimes r^1\}_0^2 = \left(\sqrt{\frac{3}{2}} \epsilon_z k_z\right) (r_0^2) , \tag{4.53}$$

$$\{\epsilon^1 \otimes k^1\}_1^2 \{r^1 \otimes r^1\}_{-1}^2 = \left(-\frac{k_z(\epsilon_x + i\epsilon_y) + (k_x + ik_y)\epsilon_z}{2}\right) (r_{-1}^2) , \tag{4.54}$$

$$\{\epsilon^1 \otimes k^1\}_{-1}^2 \{r^1 \otimes r^1\}_1^2 = \left(\frac{k_z(\epsilon_x - i\epsilon_y) + (k_x - ik_y)\epsilon_z}{2}\right) (r_1^2) , \tag{4.55}$$

$$\{\epsilon^1 \otimes k^1\}_2^2 \{r^1 \otimes r^1\}_{-2}^2 = \left(\frac{(k_x + ik_y)(\epsilon_x + i\epsilon_y)}{2}\right) (r_{-2}^2) , \tag{4.56}$$

$$\{\epsilon^1 \otimes k^1\}_{-2}^2 \{r^1 \otimes r^1\}_2^2 = \left(\frac{(k_x - ik_y)(\epsilon_x - i\epsilon_y)}{2}\right) (r_2^2) . \tag{4.57}$$

The same arguments apply for the $c = 0, 1, 2$ terms of the recoupled $\hat{T}^\dagger$ operator ending up with the values $b = 2$ and $c = 2$ for the XAS cross section. The recoupled cross section writes

$$\sigma_\omega = \pi^2 \alpha \hbar \omega k^2 \times \text{Im} \left[\sum_{a=0}^{4} (-1)^a \{\{\epsilon^{*1} \otimes k^1\}^2 \otimes \{\epsilon^1 \otimes k^1\}^2\}^a \right.$$
$$\left. \times \{\langle I | \{r^1 \otimes r^1\}^2 G^+ \{r^1 \otimes r^1\}^2 | I\rangle\}^a \right] . \tag{4.58}$$

Now one can develop (4.58) in further details for $a = 0, 1, 2, 3, 4$. We shall only report the final expression here.

4.2.3   Term $a = 0$

Let us substitute $a = 0$ in (4.58). This is the zero rank of the tensor, $\sigma(0, 0)$, describing an isotropic spectrum

$$\sigma(0, 0) = \pi^2 \alpha \hbar \omega k^2 \times \text{Im} \left[\frac{1}{10} \left(\langle I | r^2 C_{2,0}^* G^+ r^2 C_{2,0} | I\rangle + \langle I | r^2 C_{2,-1}^* G^+ r^2 C_{2,-1} | I\rangle \right. \right.$$
$$+ \langle I | r^2 C_{2,1}^* G^+ r^2 C_{2,1} | I\rangle + \langle I | r^2 C_{2,-2}^* G^+ r^2 C_{2,-2} | I\rangle$$
$$\left. \left. + \langle I | r^2 C_{2,2}^* G^+ r^2 C_{2,2} | I\rangle \right) \right] . \tag{4.59}$$

4.2.4 Term $a = 1$

The term $a = 1$ consists of three components, namely, $\sigma(1, 0)$, $\sigma(1, 1)$ and $\sigma(1, -1)$

$$
\begin{aligned}
\sigma(1, 0) = \pi^2 \alpha \hbar \omega k^2 \times \text{Im}\bigg[&\frac{i}{20}\Big((2k_x^2 + 2k_y^2 + k_z^2)\epsilon_x \epsilon_y^* - (2k_x^2 + 2k_y^2 + k_z^2)\epsilon_x^* \epsilon_y \\
&+ k_y k_z \epsilon_x \epsilon_z^* - k_y k_z \epsilon_x^* \epsilon_z + k_x k_z \epsilon_z \epsilon_y^* - k_x k_z \epsilon_y \epsilon_z^* \Big) \\
&\times \Big(\langle I | r^2 C_{2,1}^* G^+ r^2 C_{2,1} | I \rangle - \langle I | r^2 C_{2,-1}^* G^+ r^2 C_{2,-1} | I \rangle \\
&+ 2\langle I | r^2 C_{2,2}^* G^+ r^2 C_{2,2} | I \rangle - 2\langle I | r^2 C_{2,-2}^* G^+ r^2 C_{2,-2} | I \rangle \Big) \bigg],
\end{aligned}
\tag{4.60}
$$

$$
\begin{aligned}
\sigma(1, 1) = \pi^2 \alpha \hbar \omega k^2 \times \text{Im}\bigg[&\frac{1}{40}\Big((2k_x^2 + k_y^2 + 2k_z^2 - ik_x k_y)\epsilon_x^* \epsilon_z \\
&- (2k_x^2 + k_y^2 + 2k_z^2 - ik_x k_y)\epsilon_x \epsilon_z^* - i(k_x^2 + 2k_y^2 + 2k_z^2 + ik_x k_y)\epsilon_y^* \epsilon_z \\
&+ i(k_x^2 + 2k_y^2 + 2k_z^2 + ik_x k_y)\epsilon_y \epsilon_z^* + (k_y + ik_x)k_z \epsilon_x^* \epsilon_y - i(k_x - ik_y)k_z \epsilon_x \epsilon_y^* \Big) \\
&\times \Big(\sqrt{6}\langle I | r^2 C_{2,-1}^* G^+ r^2 C_{2,0} | I \rangle + \sqrt{6}\langle I | r^2 C_{2,0}^* G^+ r^2 C_{2,1} | I \rangle \\
&+ 2\langle I | r^2 C_{2,-2}^* G^+ r^2 C_{2,-1} | I \rangle + 2\langle I | r^2 C_{2,1}^* G^+ r^2 C_{2,2} | I \rangle \Big) \bigg],
\end{aligned}
\tag{4.61}
$$

$$
\begin{aligned}
\sigma(1, -1) = \pi^2 \alpha \hbar \omega k^2 \times \text{Im}\bigg[&\frac{-1}{40}\Big((2k_x^2 + k_y^2 + 2k_z^2 + ik_x k_y)\epsilon_x^* \epsilon_z \\
&- (2k_x^2 + k_y^2 + 2k_z^2 + ik_x k_y)\epsilon_x \epsilon_z^* + i(k_x^2 + 2k_y^2 + 2k_z^2 - ik_x k_y)\epsilon_y^* \epsilon_z \\
&- i(k_x^2 + 2k_y^2 + 2k_z^2 - ik_x k_y)\epsilon_y \epsilon_z^* + k_z(k_y - ik_x)\epsilon_x^* \epsilon_y + ik_z(k_x + ik_y)\epsilon_x \epsilon_y^* \Big) \\
&\times \Big(\sqrt{6}\langle I | r^2 C_{2,1}^* G^+ r_{2,0}^C | I \rangle + \sqrt{6}\langle I | r^2 C_{2,0}^* G^+ r^2 C_{2,-1} | I \rangle \\
&+ 2\langle I | r^2 C_{2,2}^* G^+ r^2 C_{2,1} | I \rangle + 2\langle I | r^2 C_{2,-1}^* G^+ r^2 C_{2,-2} | I \rangle \Big) \bigg].
\end{aligned}
\tag{4.62}
$$

A quick check of (4.60), (4.61) and (4.62) reveals that the term $a = 1$ is zero for linear light. This implies that these fundamental spectra can only be probed with circular or elliptically polarized light. It is also clear that if the conductivity tensor has no off-diagonal terms, and satisfies

$$
\begin{aligned}
\langle I | r^2 C_{2,1}^* G^+ r^2 C_{2,1} | I \rangle &= \langle I | r^2 C_{2,-1}^* G^+ r^2 C_{2,-1} | I \rangle, \\
\langle I | r^2 C_{2,2}^* G^+ r^2 C_{2,2} | I \rangle &= \langle I | r^2 C_{2,-2}^* G^+ r^2 C_{2,-2} | I \rangle,
\end{aligned}
$$

then the term $a = 1$ is again zero.

4.2.5 Term a = 2

The term $a = 2$ consists of five components, namely, $\sigma(2, 0)$, $\sigma(2, 1)$, $\sigma(2, -1)$, $\sigma(2, 2)$, and $\sigma(2, -2)$

$$
\begin{aligned}
\sigma(2, 0) = \pi^2 \alpha \hbar \omega k^2 \times \operatorname{Im}\Bigg[&\frac{1}{84}\Big((4k_x^2\epsilon_x - 2k_xk_y\epsilon_y + k_xk_z\epsilon_z + 6k_y^2\epsilon_x - 3k_z^2\epsilon_x)\epsilon_x^* \\
&+(6k_x^2\epsilon_y - 2k_xk_y\epsilon_x + 4k_y^2\epsilon_y + k_yk_z\epsilon_z - 3k_z^2\epsilon_y)\epsilon_y^* \\
&+(-3k_x^2\epsilon_z + k_xk_z\epsilon_x - 3k_y^2\epsilon_z + k_yk_z\epsilon_y - 8k_z^2\epsilon_z)\epsilon_z^* \Big) \\
&\times \Big(2\langle I|r^2 C_{2,2}^* G + r^2 C_{2,2}|I\rangle + 2\langle I|r^2 C_{2,-2}^* G + r^2 C_{2,-2}|I\rangle \\
&-2\langle I|r^2 C_{2,0}^* G + r^2 C_{2,0}|I\rangle - \langle I|r^2 C_{2,1}^* G + r^2 C_{2,1}|I\rangle \\
&-\langle I|r^2 C_{2,-1}^* G + r^2 C_{2,-1}|I\rangle \Big) \Bigg],
\end{aligned}
\tag{4.63}
$$

$$
\begin{aligned}
\sigma(2, 1) = \pi^2 \alpha \hbar \omega k^2 \times \operatorname{Im}\Bigg[&\frac{-1}{168}\Big([(4k_x\epsilon_x - 6ik_y\epsilon_x + ik_x\epsilon_y - k_y\epsilon_y)k_z \\
&+(2k_x^2 + ik_xk_y + 3k_y^2 + 2k_z^2)\epsilon_z]\epsilon_x^* - i[(-k_x\epsilon_x - ik_y\epsilon_x + 6ik_x\epsilon_y + 4k_y\epsilon_y)k_z \\
&+(3k_x^2 - ik_xk_y + 2(k_y^2 + k_z^2))\epsilon_z]\epsilon_y^* + [2k_z^2(\epsilon_x - i\epsilon_y) + k_y^2(3\epsilon_x - 2i\epsilon_y) \\
&+k_x^2(2\epsilon_x - 3i\epsilon_y) - 4ik_yk_z\epsilon_z + k_x(ik_y\epsilon_x - k_y\epsilon_y + 4k_z\epsilon_z)]\epsilon_z^* \Big) \\
&\times \Big(\sqrt{6}\langle I|r^2 C_{2,0}^* G + r^2 C_{2,1}|I\rangle - \sqrt{6}\langle I|r^2 C_{2,-1}^* G + r^2 C_{2,0}|I\rangle \\
&+6\langle I|r^2 C_{2,1}^* G + r^2 C_{2,2}|I\rangle - 6\langle I|r^2 C_{2,-2}^* G + r^2 C_{2,-1}|I\rangle \Big) \Bigg],
\end{aligned}
\tag{4.64}
$$

$$
\begin{aligned}
\sigma(2, -1) = \pi^2 \alpha \hbar \omega k^2 \times \operatorname{Im}\Bigg[&\frac{1}{168}\Big([6ik_yk_z\epsilon_x - k_yk_z\epsilon_y + 2k_x^2\epsilon_z + 3k_y^2\epsilon_z + 2k_z^2\epsilon_z \\
&-ik_x(4ik_z\epsilon_x + k_z\epsilon_y + k_y\epsilon_z)]\epsilon_x^* + i[-k_xk_z(\epsilon_x + 6i\epsilon_y) + k_yk_z(i\epsilon_x + 4\epsilon_y) \\
&+3k_x^2\epsilon_z + ik_xk_y\epsilon_z + 2(k_y^2 + kz^2)\epsilon_z]\epsilon_y^* + [2k_z^2(\epsilon_x + i\epsilon_y) + k_y^2(3\epsilon_x + 2i\epsilon_y) \\
&+k_x^2(2\epsilon_x + 3i\epsilon_y) + 4ik_ykz\epsilon_z + k_x(-ik_y\epsilon_x - k_y\epsilon_y + 4k_z\epsilon_z)]\epsilon_z^* \Big) \\
&\times \Big(\sqrt{6}\langle I|r^2 C_{2,0}^* G + r^2 C_{2,-1}|I\rangle - \sqrt{6}\langle I|r^2 C_{2,1}^* G + r^2 C_{2,0}|I\rangle \\
&+6\langle I|r^2 C_{2,-1}^* G + r^2 C_{2,-2}|I\rangle - 6\langle I|r^2 C_{2,2}^* G + r^2 C_{2,1}|I\rangle \Big) \Bigg],
\end{aligned}
\tag{4.65}
$$

$$\sigma(2, 2) = \pi^2\alpha\hbar\omega k^2 \times \mathrm{Im}\left[\frac{1}{4\sqrt{21}}\Big(-2(k_x - ik_y)(\epsilon_x^* - i\epsilon_y^*)(k_x\epsilon_x + k_y\epsilon_y - 2k_z\epsilon_z)\right.$$

$$-3(\epsilon_z(k_x - ik_y) + k_z(\epsilon_x - i\epsilon_y))(\epsilon_z^*(k_x - ik_y) + k_z\epsilon_x^* - ik_z\epsilon_y^*)$$

$$-2(k_x - ik_y)(\epsilon_x - i\epsilon_y)(k_x\epsilon_x^* + k_y\epsilon_y^* - 2k_z\epsilon_z^*)\Big)$$

$$\times\left(\sqrt{\frac{2}{7}}\langle I|r^2 C_{2,0}^* G^+ r^2 C_{2,2}|I\rangle + \sqrt{\frac{3}{7}}\langle I|r^2 C_{2,-1}^* G^+ r^2 C_{2,1}|I\rangle\right.$$

$$\left.\left.+\sqrt{\frac{2}{7}}\langle I|r^2 C_{2,-2}^* G^+ r^2 C_{2,0}|I\rangle\right)\right], \tag{4.66}$$

$$\sigma(2, -2) = \pi^2\alpha\hbar\omega k^2 \times \mathrm{Im}\left[\frac{1}{4\sqrt{21}}\Big(-2(k_x + ik_y)(\epsilon_x^* + i\epsilon_y^*)(k_x\epsilon_x + k_y\epsilon_y - 2k_z\epsilon_z)\right.$$

$$-3(\epsilon_z(k_x + ik_y) + k_z(\epsilon_x + i\epsilon_y))(\epsilon_z^*(k_x + ik_y) + k_z\epsilon_x^* + ik_z\epsilon_y^*)$$

$$-2(k_x + ik_y)(\epsilon_x + i\epsilon_y)(k_x\epsilon_x^* + k_y\epsilon_y^* - 2k_z\epsilon_z^*)\Big)$$

$$\times\left(\sqrt{\frac{2}{7}}\langle I|r^2 C_{2,0}^* G^+ r^2 C_{2,-2}|I\rangle + \sqrt{\frac{3}{7}}\langle I|r^2 C_{2,1}^* G^+ r^2 C_{2,-1}|I\rangle\right.$$

$$\left.\left.+\sqrt{\frac{2}{7}}\langle I|r^2 C_{2,2}^* G^+ r^2 C_{2,0}|I\rangle\right)\right]. \tag{4.67}$$

4.2.6 Term a = 3

The term $a = 3$ consists of seven components, namely, $\sigma(3, 0)$, $\sigma(3, 1)$, $\sigma(3, -1)$, $\sigma(3, 2)$, $\sigma(3, -2)$, $\sigma(3, 3)$, and $\sigma(3, -3)$

$$\sigma(3, 0) = \pi^2\alpha\hbar\omega k^2 \times \mathrm{Im}\left[\frac{i}{20}\Big((k_x^2 + k_y^2 - 2k_z^2)\epsilon_x\epsilon_y^* - (k_x^2 + k_y^2 - 2k_z^2)\epsilon_y\epsilon_x^*\right.$$

$$+2k_xk_z\epsilon_y\epsilon_z^* - 2k_xk_z\epsilon_z\epsilon_y^* + 2k_yk_z\epsilon_z\epsilon_x^* - 2k_yk_z\epsilon_x\epsilon_z^*\Big)$$

$$\times\Big(2\langle I|r^2 C_{2,-1}^* G^+ r^2 C_{2,-1}|I\rangle - 2\langle I|r^2 C_{2,1}^* G^+ r^2 C_{2,1}|I\rangle$$

$$\left.+\langle I|r^2 C_{2,2}^* G^+ r^2 C_{2,2}|I\rangle - \langle I|r^2 C_{2,-2}^* G^+ r^2 C_{2,-2}|I\rangle\Big)\right], \tag{4.68}$$

$$\sigma(3,1) = \pi^2 \alpha \hbar \omega k^2 \times \text{Im}\Bigg[\frac{1}{40\sqrt{6}}\Big(i(3k_x^2 - 2ik_x k_y + k_y^2 - 4k_z^2)\epsilon_z \epsilon_y^*$$

$$+8k_z(k_y + ik_x)\epsilon_x \epsilon_y^* - (k_x^2 + 2ik_x k_y + 3k_y^2 - 4k_z^2)\epsilon_x^* \epsilon_z - 8k_z(k_y + ik_x)\epsilon_y \epsilon_x^*$$

$$+\epsilon_z^*[k_x^2(\epsilon_x - 3i\epsilon_y) + 2ik_x k_y(\epsilon_x + i\epsilon_y) + k_y^2(3\epsilon_x - i\epsilon_y) - 4k_z^2(\epsilon_x - i\epsilon_y)]\Big)$$

$$\times\Big(2\langle I|r^2 C_{2,0}^* G^+ r^2 C_{2,1}|I\rangle + 2\langle I|r^2 C_{2,-1}^* G^+ r^2 C_{2,0}|I\rangle$$

$$-\sqrt{6}\langle I|r^2 C_{2,1}^* G^+ r^2 C_{2,2}|I\rangle - \sqrt{6}\langle I|r^2 C_{2,-2}^* G^+ r^2 C_{2,-1}|I\rangle\Big)\Bigg], \qquad (4.69)$$

$$\sigma(3,-1) = \pi^2 \alpha \hbar \omega k^2 \times \Im\Bigg[\frac{1}{40\sqrt{6}}\Big(i(3k_x^2 + 2ik_x k_y + k_y^2 - 4k_z^2)\epsilon_y^* \epsilon_z$$

$$+8ik_z(k_x + ik_y)\epsilon_x \epsilon_y^* + (k_x^2 - 2ik_x k_y + 3k_y^2 - 4k_z^2)\epsilon_x^* \epsilon_z + 8k_z(k_y - ik_x)\epsilon_x^* \epsilon_y$$

$$+\epsilon_z^*[k_x^2(-(\epsilon_x + 3i\epsilon_y)) + 2k_x k_y(\epsilon_y + i\epsilon_x) - k_y^2(3\epsilon_x + i\epsilon_y) + 4k_z^2(\epsilon_x + i\epsilon_y)]\Big)$$

$$\times\Big(2\langle I|r^2 C_{2,0}^* G^+ r^2 C_{2,-1}|I\rangle + 2\langle I|r^2 C_{2,1}^* G^+ r^2 C_{2,0}|I\rangle$$

$$-\sqrt{6}\langle I|r^2 C_{2,-1}^* G^+ r^2 C_{2,-2}|I\rangle - \sqrt{6}\langle I|r^2 C_{2,2}^* G^+ r^2 C_{2,1}|I\rangle\Big)\Bigg], \qquad (4.70)$$

$$\sigma(3,2) = \pi^2 \alpha \hbar \omega k^2 \times \text{Im}\Bigg[\frac{1}{4\sqrt{6}}\Big(k_y + ik_x\Big)\Big(\epsilon_x^*(k_x \epsilon_y - ik_y \epsilon_y + 2ik_z \epsilon_z)$$

$$+\epsilon_y^*(-k_x \epsilon_x + ik_y \epsilon_x + 2k_z \epsilon_z) - 2ik_z \epsilon_z^*(\epsilon_x - i\epsilon_y)\Big)$$

$$\times\Big(\langle I|r^2 C_{2,0}^* G^+ r^2 C_{2,2}|I\rangle - \langle I|r^2 C_{2,-2}^* G^+ r^2 C_{2,0}|I\rangle\Big)\Bigg], \qquad (4.71)$$

$$\sigma(3,-2) = \pi^2 \alpha \hbar \omega k^2 \times \text{Im}\Bigg[\frac{1}{4\sqrt{6}}\Big(ik_y + k_x\Big)\Big(\epsilon_x^*(ik_x \epsilon_y - k_y \epsilon_y + 2k_z \epsilon_z)$$

$$+\epsilon_y^*(-ik_x \epsilon_x + k_y \epsilon_x + 2ik_z \epsilon_z) - 2k_z \epsilon_z^*(\epsilon_x + i\epsilon_y)\Big)$$

$$\times\Big(\langle I|r^2 C_{2,2}^* G^+ r^2 C_{2,0}|I\rangle - \langle I|r^2 C_{2,0}^* G^+ r^2 C_{2,-2}|I\rangle\Big)\Bigg], \qquad (4.72)$$

$$\sigma(3,3) = \pi^2 \alpha \hbar \omega k^2 \times \text{Im}\Bigg[\frac{1}{8}\Big((k_x - ik_y)^2\Big)\Big(-\epsilon_z^*(\epsilon_x - i\epsilon_y) + \epsilon_z(\epsilon_x^* - i\epsilon_y^*)\Big)$$

$$\times\Big(-\langle I|r^2 C_{2,-1}^* G^+ r^2 C_{2,2}|I\rangle - \langle I|r^2 C_{2,-2}^* G^+ r^2 C_{2,1}|I\rangle\Big)\Bigg], \qquad (4.73)$$

$$\sigma(3, -3) = \pi^2 \alpha \hbar \omega k^2 \times \text{Im} \left[\frac{1}{8} \left((k_x + ik_y)^2 \right) \left(-\epsilon_z^* (\epsilon_x + i\epsilon_y) + \epsilon_z (\epsilon_x^* + i\epsilon_y^*) \right) \right.$$

$$\left. \times \left(\langle I | r^2 C_{2,1}^* G^+ r^2 C_{2,-2} | I \rangle + \langle I | r^2 C_{2,2}^* G^+ r^2 C_{2,-1} | I \rangle \right) \right]. \quad (4.74)$$

4.2.7 Term a = 4

The term $a = 4$ consists of nine components, namely, $\sigma(4, 0)$, $\sigma(4, 1)$, $\sigma(4, -1)$, $\sigma(4, 2)$, $\sigma(4, -2)$, $\sigma(4, 3)$, $\sigma(4, -3)$, $\sigma(4, 4)$, and $\sigma(4, -4)$

$$\sigma(4, 0) = \pi^2 \alpha \hbar \omega k^2 \times \text{Im} \left[\frac{1}{140} \left([3k_x^2 \epsilon_x + k_y^2 \epsilon_x - 4k_z^2 \epsilon_x + 2k_x k_y \epsilon_y - 8k_x k_z \epsilon_z] \epsilon_x^* \right. \right.$$

$$+ [k_x^2 \epsilon_y + 3k_y^2 \epsilon_y - 4k_z^2 \epsilon_y + 2k_x k_y \epsilon_x - 8k_y k_z \epsilon_z] \epsilon_y^*$$

$$\left. - 4\epsilon_z^* [\epsilon_z (k_x^2 + k_y^2 - 2k_z^2) + 2k_z (k_x \epsilon_x + k_y \epsilon_y)] \right)$$

$$\times \left(6 \langle I | r^2 C_{2,0}^* G^+ r^2 C_{2,0} | I \rangle - 4 \langle I | r^2 C_{2,1}^* G^+ r^2 C_{2,1} | I \rangle \right.$$

$$- 4 \langle I | r^2 C_{2,-1}^* G^+ r^2 C_{2,-1} | I \rangle + \langle I | r^2 C_{2,2}^* G^+ r^2 C_{2,2} | I \rangle$$

$$\left. \left. + \langle I | r^2 C_{2,-2}^* G^+ r^2 C_{2,-2} | I \rangle \right) \right], \quad (4.75)$$

$$\sigma(4, 1) = \pi^2 \alpha \hbar \omega k^2 \times \text{Im} \left[\frac{1}{56} \left(\epsilon_x^* [3k_z^2 \epsilon_z - 2ik_x (k_y \epsilon_z + 3ik_z \epsilon_x + k_z \epsilon_y) + \epsilon_z (k_y^2 - 4k_z^2) \right. \right.$$

$$+ 2k_y k_z (\epsilon_y - i\epsilon_x)] - i\epsilon_y^* [\epsilon_z (k_x^2 + 2ik_x k_y + 3k_y^2 - 4k_z^2)$$

$$+ 2k_z (k_x \epsilon_x + ik_x \epsilon_y + ik_y \epsilon_x + 3k_y \epsilon_y)] + \epsilon_z^* [k_x^2 (3\epsilon_x - i\epsilon_y)$$

$$\left. + 2k_x k_y (\epsilon_y - i\epsilon_x) - 8k_x k_z \epsilon_z + k_y^2 (\epsilon_x - 3i\epsilon_y) + 8ik_y k_z \epsilon_z - 4k_z^2 (\epsilon_x - i\epsilon_y)] \right)$$

$$\times \left(\sqrt{6} \langle I | r^2 C_{2,0}^* G^+ r^2 C_{2,1} | I \rangle - \sqrt{6} \langle I | r^2 C_{2,-1}^* G^+ r^2 C_{2,0} | I \rangle \right.$$

$$\left. \left. \langle I | r^2 C_{2,-2}^* G^+ r^2 C_{2,-1} | I \rangle - \langle I | r^2 C_{2,1}^* G^+ r^2 C_{2,2} | I \rangle \right) \right], \quad (4.76)$$

$$\sigma(4,-1) = \pi^2 \alpha \hbar \omega k^2 \times \text{Im}\left[\frac{-1}{56}\left(\epsilon_x^*[\epsilon_z(3k_x^2 + 2ik_xk_y + k_y^2 - 4k_z^2)\right.\right.$$
$$+2k_z(3k_x\epsilon_x + ik_x\epsilon_y + ik_y\epsilon_x + k_y\epsilon_y)] + \epsilon_y^*[i\epsilon_z(k_x^2 - 2ik_xk_y + 3k_y^2 - 4k_z^2)$$
$$+2k_z(ik_x\epsilon_x + k_x\epsilon_y + k_y\epsilon_x + 3ik_y\epsilon_y)] + \epsilon_z^*[k_x^2(3\epsilon_x + i\epsilon_y)$$
$$+2k_xk_y(\epsilon_y + i\epsilon_x) - 8k_xk_z\epsilon_z + k_y^2(\epsilon_x + 3i\epsilon_y) - 8ik_yk_z\epsilon_z - 4k_z^2(\epsilon_x + i\epsilon_y)]\Big)$$
$$\times\left(\sqrt{6}\langle I|r^2 C_{2,0}^* G^+ r^2 C_{2,-1}|I\rangle - \sqrt{6}\langle I|r^2 C_{2,1}^* G^+ r^2 C_{2,0}|I\rangle\right.$$
$$\left.\left.\langle I|r^2 C_{2,2}^* G^+ r^2 C_{2,1}|I\rangle - \langle I|r^2 C_{2,-1}^* G^+ r^2 C_{2,-2}|I\rangle\right)\right], \tag{4.77}$$

$$\sigma(4,2) = \pi^2 \alpha \hbar \omega k^2 \times \text{Im}\left[\frac{1}{56}\left(-(k_x - ik_y)(\epsilon_x^* - i\epsilon_y^*)(k_x\epsilon_x + k_y\epsilon_y - 2k_z\epsilon_z)\right.\right.$$
$$+2[\epsilon_z(k_x - ik_y) + k_z(\epsilon_x - i\epsilon_y)][\epsilon_z^*(k_x - ik_y) + k_z\epsilon_x^* - ik_z\epsilon_y^*]$$
$$-(k_x - ik_y)(\epsilon_x - i\epsilon_y)(k_x\epsilon_x^* + k_y\epsilon_y^* - 2k_z\epsilon_z^*)\Big)$$
$$\times\left(\sqrt{6}\langle I|r^2 C_{2,0}^* G^+ r^2 C_{2,2}|I\rangle + \sqrt{6}\langle I|r^2 C_{2,-2}^* G^+ r^2 C_{2,0}|I\rangle\right.$$
$$\left.\left.-4\langle I|r^2 C_{2,-1}^* G^+ r^2 C_{2,1}|I\rangle\right)\right], \tag{4.78}$$

$$\sigma(4,-2) = \pi^2 \alpha \hbar \omega k^2 \times \text{Im}\left[\frac{1}{56}\left(-(k_x + ik_y)(\epsilon_x^* + i\epsilon_y^*)(k_x\epsilon_x + k_y\epsilon_y - 2k_z\epsilon_z)\right.\right.$$
$$+2[\epsilon_z(k_x + ik_y) + k_z(\epsilon_x + i\epsilon_y)][\epsilon_z^*(k_x + ik_y) + k_z\epsilon_x^* + ik_z\epsilon_y^*]$$
$$-(k_x + ik_y)(\epsilon_x + i\epsilon_y)(k_x\epsilon_x^* + k_y\epsilon_y^* - 2k_z\epsilon_z^*)\Big)$$
$$\times\left(\sqrt{6}\langle I|r^2 C_{2,0}^* G^+ r^2 C_{2,-2}|I\rangle + \sqrt{6}\langle I|r^2 C_{2,2}^* G^+ r^2 C_{2,0}|I\rangle\right.$$
$$\left.\left.-4\langle I|r^2 C_{2,1}^* G^+ r^2 C_{2,-1}|I\rangle\right)\right], \tag{4.79}$$

$$\sigma(4,3) = \pi^2 \alpha \hbar \omega k^2 \times \text{Im}\left[\frac{-1}{8}(k_x - ik_y)\left((\epsilon_x^* - i\epsilon_y^*)(\epsilon_z(k_x - ik_y)\right.\right.$$
$$+2k_z(\epsilon_x - i\epsilon_y)) + \epsilon_z^*(k_x - ik_y)(\epsilon_x - i\epsilon_y)\Big)$$
$$\times\left(\langle I|r^2 C_{2,-2}^* G^+ r^2 C_{2,1}|I\rangle - \langle I|r^2 C_{2,-1}^* G^+ r^2 C_{2,2}|I\rangle\right)\right], \tag{4.80}$$

$$\sigma(4, -3) = \pi^2 \alpha \hbar \omega k^2 \times \text{Im} \left[\frac{1}{8} \left(k_x + i k_y \right) \left((\epsilon_x^* + i \epsilon_y^*)(\epsilon_z (k_x + i k_y) \right. \right.$$

$$+ 2k_z(\epsilon_x + i\epsilon_y)) + \epsilon_z^*(k_x + ik_y)(\epsilon_x + i\epsilon_y) \Big)$$

$$\times \left(\langle I | r^2 C_{2,2}^* G^+ r^2 C_{2,-1} | I \rangle - \langle I | r^2 C_{2,1}^* G^+ r^2 C_{2,-2} | I \rangle \right) \Bigg], \quad (4.81)$$

$$\sigma(4, 4) = \pi^2 \alpha \hbar \omega k^2 \times \text{Im} \left[\frac{1}{4} \left((k_x - i k_y)^2 (\epsilon_x - i\epsilon_y)(\epsilon_x^* - i\epsilon_y^*) \right) \right.$$

$$\times \left(\langle I | r^2 C_{2,-2}^* G^+ r^2 C_{2,2} | I \rangle \right) \Bigg], \quad (4.82)$$

$$\sigma(4, -4) = \pi^2 \alpha \hbar \omega k^2 \times \text{Im} \left[\frac{1}{4} \left((k_x + i k_y)^2 (\epsilon_x + i\epsilon_y)(\epsilon_x^* + i\epsilon_y^*) \right) \right.$$

$$\times \left(\langle I | r^2 C_{2,2}^* G^+ r^2 C_{2,-2} | I \rangle \right) \Bigg]. \quad (4.83)$$

In the most general case, the quadrupole XAS signal can be described using 25 fundamental spectra as given in (4.59)–(4.83). Although the expression seems at first sight complicated, major simplifications and intuitive conclusions can be made when one considers the symmetry of the absorbing system. We shall illustrate this in the following section.

4.2.7.1 Case Study of a d^9 Ion

As an example, we will study again a d^9 ion in different local symmetries.

Spherical Symmetry

We shall start with an isolated d^9 ion (i.e., spherical symmetry). The conductivity tensor of such an ion is shown in Fig. 4.11. The tensor consists of 25 elements that form the 25 fundamental spectra through appropriate linear combinations. Only the five diagonal elements are non-zero in this case and are all equal. This means that the only possibly active fundamental spectra are of the type $\sigma(a, 0)$ with $a = 0, 1, 2, 3, 4$. However, because all the diagonal elements are equal, only the fundamental spectrum $\sigma(0, 0)$ is non-zero. This fundamental spectrum has no angular dependence, hence this system is isotropic. It is not a surprising result that for a spherical system, no angular dependence would be observed.

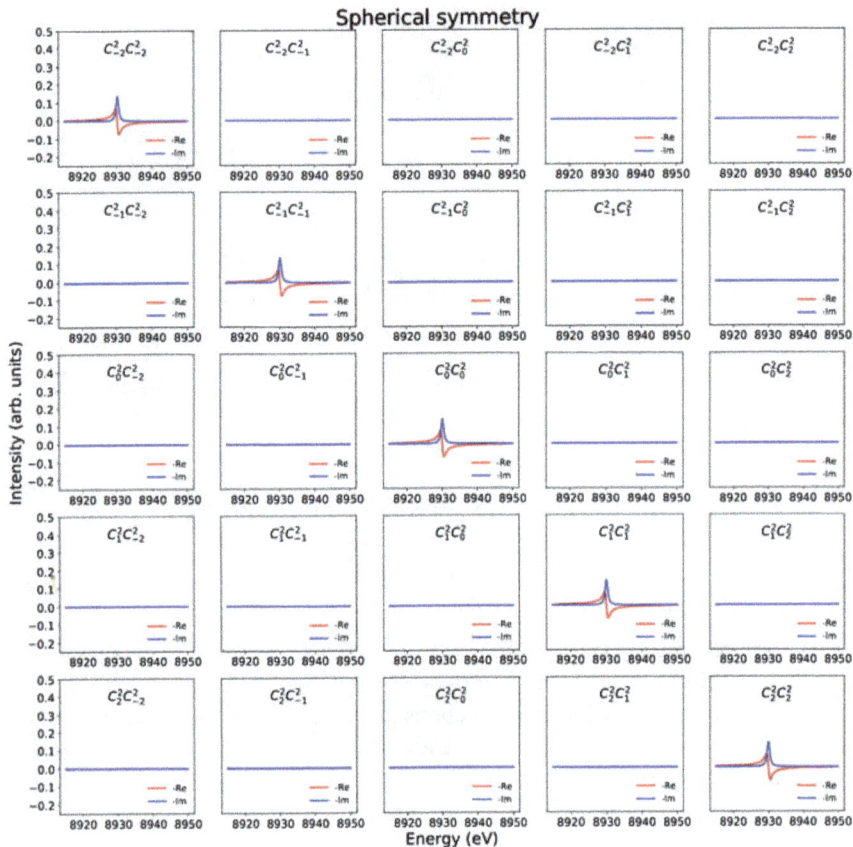

Fig. 4.11 Conductivity tensor calculated for a d^9 ion in spherical symmetry. Re and Im are the real and imaginary parts of the tensor

Octahedral Crystal Field

We shall examine next a d^9 ion in O_h symmetry. The conductivity tensor of this ion is shown in Fig. 4.12. Several differences can be directly seen in comparison with the previous case:

- The elements with the transition operator $T = C_{-2}^2$ are mixed with those of $T = C_2^2$.
- The diagonal elements are not equal.

Let us consider the first point. This mixing leads to the same form of eigenvectors than for obtained for the $3d$ orbitals ($Y_{2,m}$) for an O_h crystal field [see (4.23)]. Indeed, this is exactly the same problem. In order to obtain only diagonal elements, one can apply the following rotation (4.84):

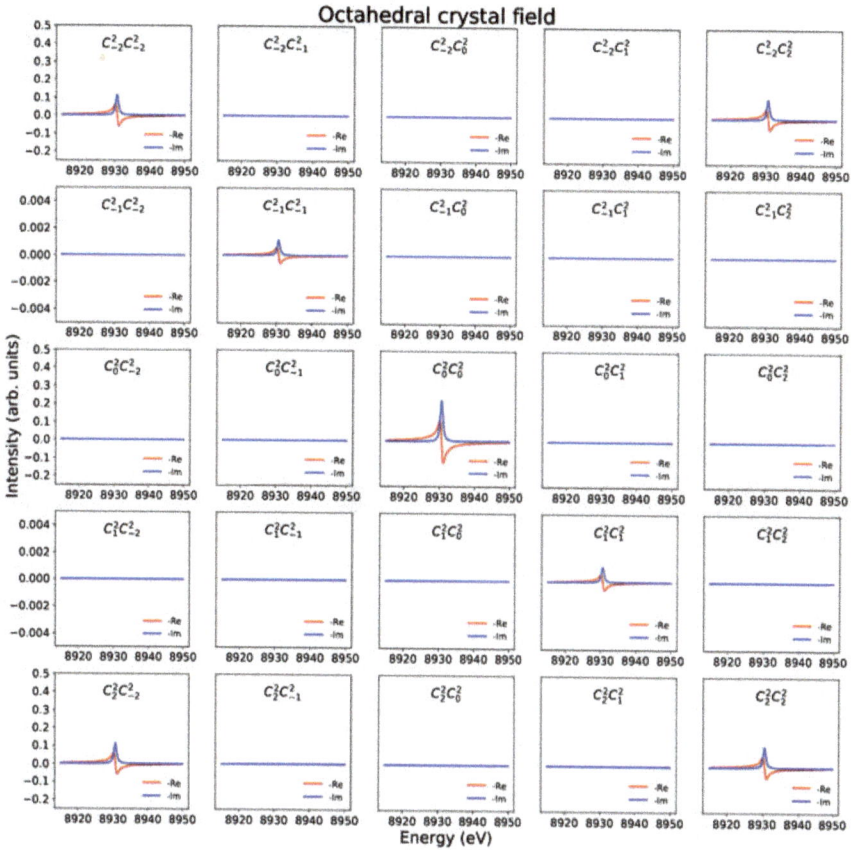

Fig. 4.12 Conductivity tensor of a d^9 ion in an octahedral crystal field ($10D_q = 1.1$ eV). Re and Im are the real and imaginary parts of the tensor

$$
Rot = \begin{bmatrix}
\frac{1}{\sqrt{2}} & 0 & 0 & 0 & \frac{1}{\sqrt{2}} \\
0 & 0 & 1 & 0 & 0 \\
0 & \frac{i}{\sqrt{2}} & 0 & \frac{i}{\sqrt{2}} & 0 \\
0 & \frac{1}{\sqrt{2}} & 0 & -\frac{1}{\sqrt{2}} & 1 \\
\frac{i}{\sqrt{2}} & 0 & 0 & 0 & -\frac{i}{\sqrt{2}}
\end{bmatrix}.
\tag{4.84}
$$

The rotated conductivity tensor is shown in Fig. 4.13. Only diagonal elements exist now. These are separated into two types: three (t_{2g}) that are equal with transition operators, C_{xy}^2, C_{yz}^2 and C_{xz}^2, and two (e_g) that are equal with transition operators, $C_{z^2}^2$ and $C_{x^2-y^2}^2$.

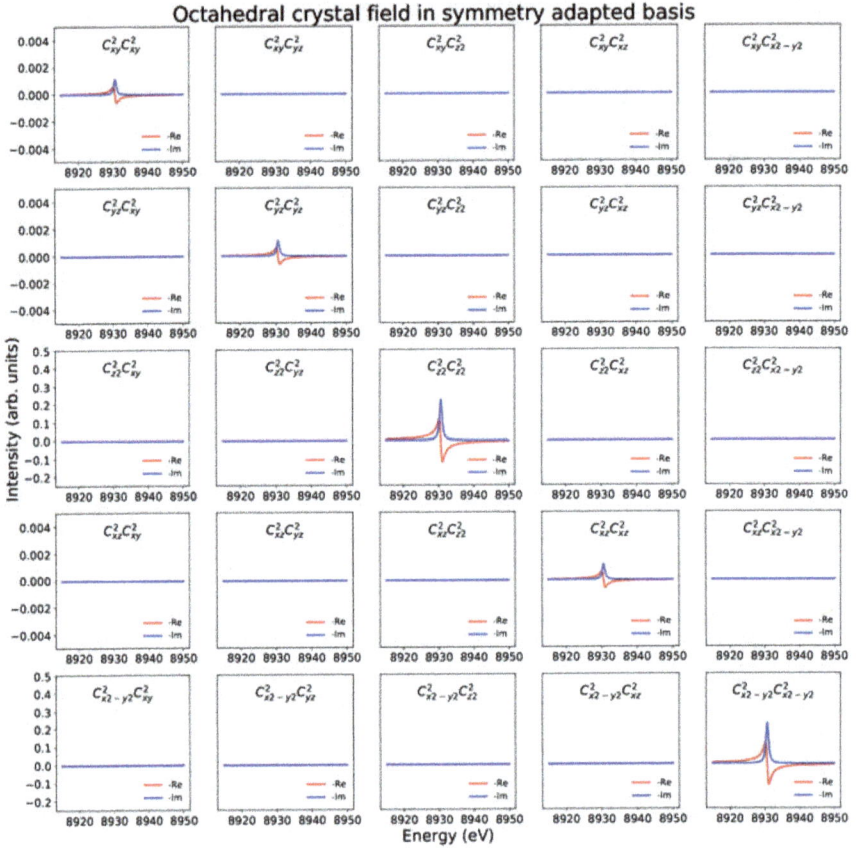

Fig. 4.13 Conductivity tensor for a d^9 ion in an octahedral crystal field ($10D_q = 1.1$ eV) calculated using the symmetry adapted basis set

Five fundamental spectra come into play, namely, $\sigma(0, 0)$, $\sigma(2, 0)$, $\sigma(4, 0)$, $\sigma(4, 4)$, and $\sigma(4, -4)$ as shown in Fig. 4.14. The O_h symmetry implies that $R(2, 0)$ is always equal to zero as confirmed by the calculation. In addition, $R(4, 4) = R(4, -4)$ and are proportional to $R(4, 0)$ as can be seen from (4.75), (4.82), and (4.83). Therefore, as can be expected from group theory, only two fundamental spectra are required to fully describe the system.

Let us investigate the angular dependence of a quadrupole transition in an O_h crystal field considering two scattering geometries. In the first geometry, the wave vector (k) is aligned parallel to the [100] direction and the polarization (ϵ) is rotated in the $z - y$-plane as illustrated in the right panel of Fig. 4.16. Despite the presence of non-isotropic fundamental spectra [$\sigma(4, 0)$, $\sigma(4, 4)$, and $\sigma(4, -4)$], the XAS cross-section is constant in these settings as shown in Fig. 4.16a. In the second geometry we

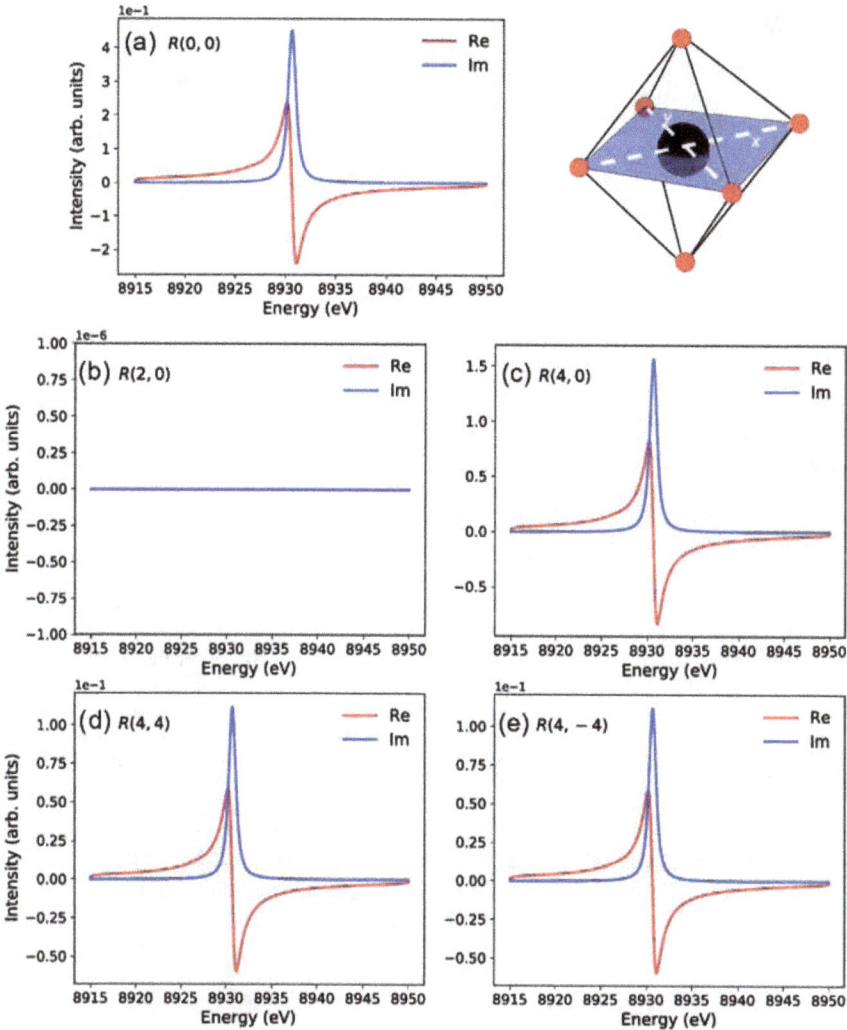

Fig. 4.14 Fundamental spectra for a quadrupole transition in a $3d^9$ ion in an octahedral crystal field. Re and Im are the real and imaginary parts of the spectra

have $k \parallel [\frac{1}{\sqrt{2}} \frac{1}{\sqrt{2}} 0]$ and ϵ is rotated about the $[\frac{1}{\sqrt{2}} \frac{1}{\sqrt{2}} 0]$ axis as depicted in Fig. 4.16b. The XAS cross section shows a clear twofold angular dependence in these settings.

It is interesting to discuss the difference between both scattering geometries and the reason behind the absence of angular dependence in the first case. In Fig. 4.15, we plot the angular dependence of the light tensor $[E(4, 0), E(4, 4), \text{and } E(4, -4)]$ for both cases. The contribution of the term $E(4, 0)$ is 90° out-of-phase with respect to the terms $E(4, 4)$ and $E(4, -4)$ for the first scattering geometry [see panel (a) of Fig. 4.15]. The O_h symmetry implies that the ratio between the $R(4, 0)$ and the

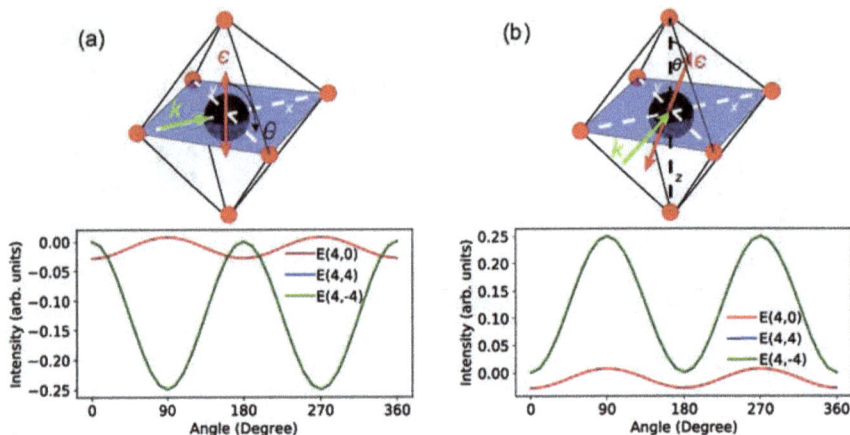

Fig. 4.15 Angular dependence of $E(4, 0)$, $E(4, 4)$, and $E(4, -4)$ terms. **a** The wave vector (k) is aligned with [100]. The angular dependence is computed by rotating the polarization (ϵ) about [100] with $\theta = 0°$ for $\epsilon \parallel$ [001]. **b** k is aligned with [$\frac{1}{\sqrt{2}}\frac{1}{\sqrt{2}}0$]. The angular dependence is computed by rotating ϵ about [$\frac{1}{\sqrt{2}}\frac{1}{\sqrt{2}}0$] with $\theta = 0°$ for $\epsilon \parallel$ [001]

$R(4, \pm4)$ terms leads to a constant XAS cross section. On the other hand, as depicted in Fig. 4.15b, all terms are in phase which leads to an angular dependent XAS.

An important distinction between the dipole and quadrupole transitions can be concluded from these examples. While a dipole transition in an O_h system exhibits no angular dependence, the quadrupole transition can show angular dependences when the scattering geometry is appropriately chosen. This difference holds because a quadrupole transition has higher multipole contributions that give rise to angular dependences not observable for dipole transitions.

Tetragonal Crystal Field

The effect of reducing the crystal field symmetry to tetragonal by applying a compressive distortion along the z-axis can be directly seen in the angular dependence of the quadrupole transition. In contrast to the case of O_h crystal field (see Fig. 4.16a), now rotating ϵ about the [100] axis shows angular dependence because the z- and y-axes are not equivalent (see Fig. 4.17a). However, as could be expected, rotating ϵ about the [001] axis shows no angular dependence (see Fig. 4.17b). In this projection, the system is effective of O_h symmetry.

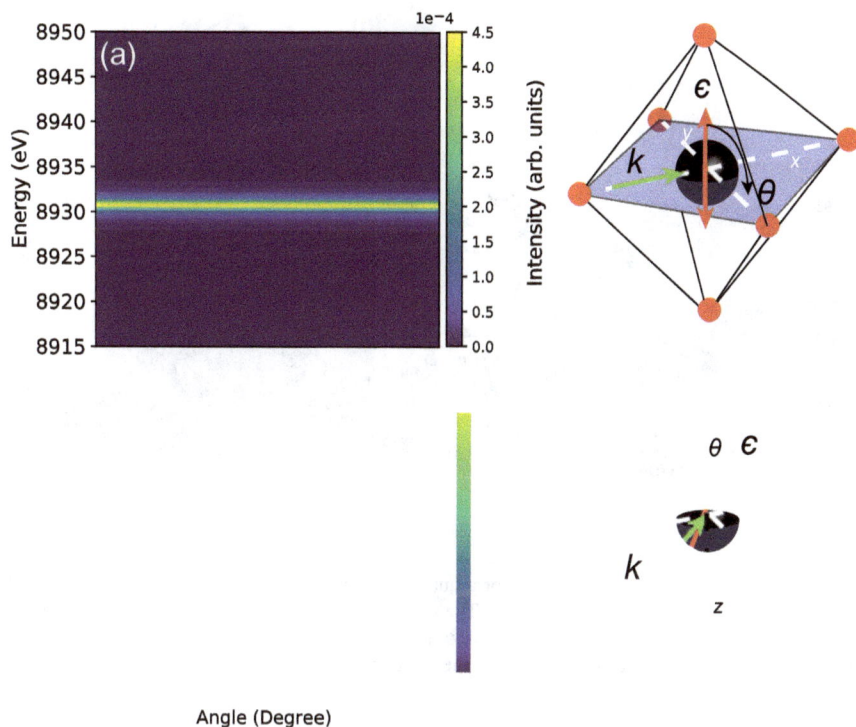

Fig. 4.16 Angular dependence of XAS for a quadrupole transition in a $3d^9$ O_h ion with $10D_q = 1.1$ eV. **a** The wave vector (k) is aligned with [100]. The angular dependence is computed by rotating the polarization vector (ϵ) about [100] with $\theta = 0°$ for $\epsilon \parallel$ [001] as depicted in the sketch on the right. **b** k is aligned with $[\frac{1}{\sqrt{2}} \frac{1}{\sqrt{2}} 0]$. The angular dependence is computed by rotating ϵ about $[\frac{1}{\sqrt{2}} \frac{1}{\sqrt{2}} 0]$ with $\theta = 0°$ for $\epsilon \parallel$ [001] as depicted in the sketch on the right

Octahedral Crystal Field with Exchange Field $\parallel$ z

Consider a magnetic $3d^9$ ion where the crystal field is O_h with an exchange field aligned along the z-axis. Seven fundamental spectra come into play, namely, $R(0, 0)$, $R(1, 0)$, $R(2, 0)$, $R(3, 0)$, $R(4, 0)$, $R(4, -4)$, and $R(4, 4)$. We have shown previously that for O_h symmetry, when k is aligned parallel to the [100] direction, and ϵ is rotated in the $z - y$-plane, angular dependence is observed (see Fig. 4.16a). Repeating the same calculation with an exchange field aligned along the z-axis leads to an angular dependent XAS as shown in Fig. 4.18a. The exchange field reduces the symmetry along the z-axis. The effects of rotating the incident linear polarization in the $z - y$-plane on the fundamental cross sections are shown in Fig. 4.19. Only the terms $\sigma(2, 0)$, $\sigma(4, 0)$, $\sigma(4, 4)$, and $\sigma(4, -4)$ are non-zero and exhibit a twofold angular

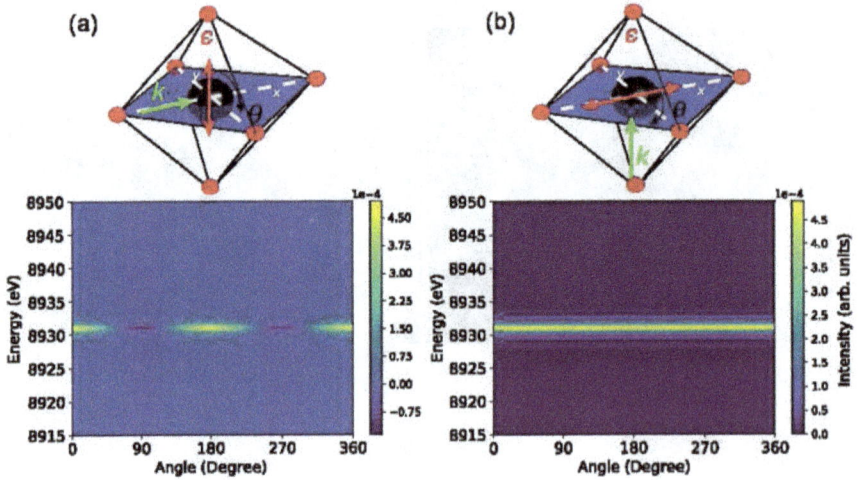

Fig. 4.17 Angular dependence of XAS for a quadrupole transition in a $3d^9$ ion in a D_{4h} crystal field ($10D_q = 1.1$ eV and $D_s = -0.2$ eV). **a** The wave vector (k) is aligned with [100]. The angular dependence is computed by rotating the polarization (ϵ) about [100] with $\theta = 0°$ for $\epsilon \parallel$ [001]. **b** k is aligned with [001] and ϵ is rotated about the [001] with $\theta = 0°$ for $\epsilon \parallel$ [100]

dependence. However, one notes that $\sigma(4, 0)$ is 90° shifted with respect to $\sigma(4, 4)$ and $\sigma(4, -4)$ which implies that the angular dependence of the XAS will be small. In comparison, no angular dependence is observed when ϵ is rotated in the $x - y$-plane as shown in Fig. 4.18b.

Finally, the exchange field can give rise to interesting combinations of structural and magnetic dichroism effects. Consider aligning $k \parallel$ [001] and measuring XAS using circular polarized light. Rotating the system about the [100] axis gives rise to unconventional angular dependent XAS as shown in Fig. 4.20. This angular dependence arises from a combination of structural and magnetic dichroism effects.

The magnetic contribution arises from the circular dichroism active terms which are $\sigma(1, 0)$ and $\sigma(3, 0)$ (see Fig. 4.21a). On the other hand, the structural contribution arises from the linear dichroism active terms which are $\sigma(4, 0)$, $\sigma(4, 4)$, and $\sigma(4, -4)$ (see Fig. 4.21b). In addition, these terms contribute weakly to the magnetic dichroism.

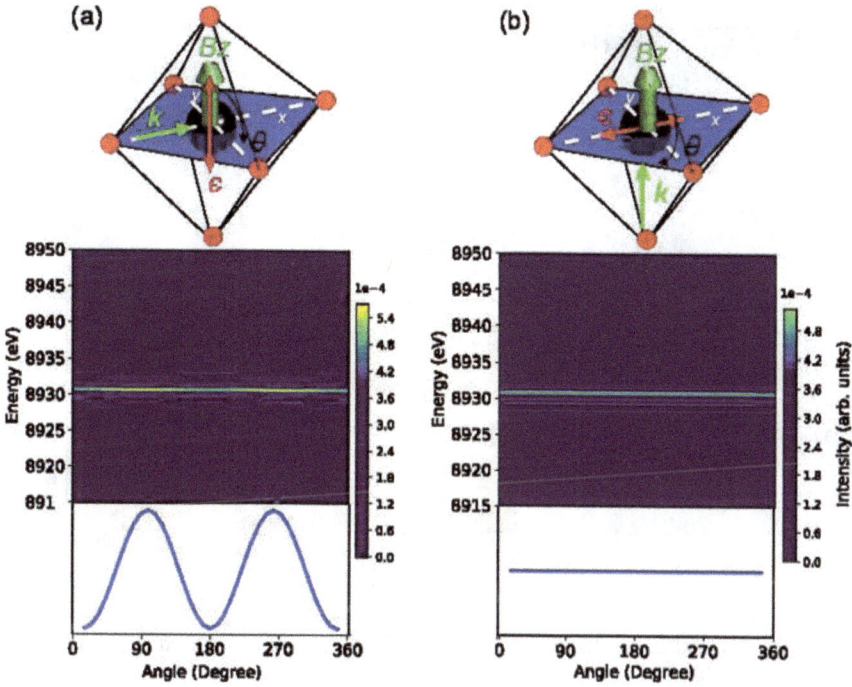

Fig. 4.18 Angular dependence of a quadrupole transition for a $3d^9$ ion in O_h crystal field ($10D_q = 1.1\,\text{eV}$) with an exchange field aligned along the z-axis ($B = 0.05\,\text{eV}$). **a** The wave vector (k) is aligned with [100]. The angular dependence is computed by rotating the polarization vector (ϵ) about [100] with $\theta = 0°$ for $\epsilon \parallel$ [001]. **b** k is aligned with [001] and ϵ is rotated about the [001] with $\theta = 0°$ for $\epsilon \parallel$ [100]

This is illustrated in Fig. 4.22 where in panel (a) we show the structural dichroism signal for the same system without an exchange field and in panel (b) the difference between the case with exchange versus without exchange. The magnetic contribution is about three orders of magnitude less than the structural contribution for this system.

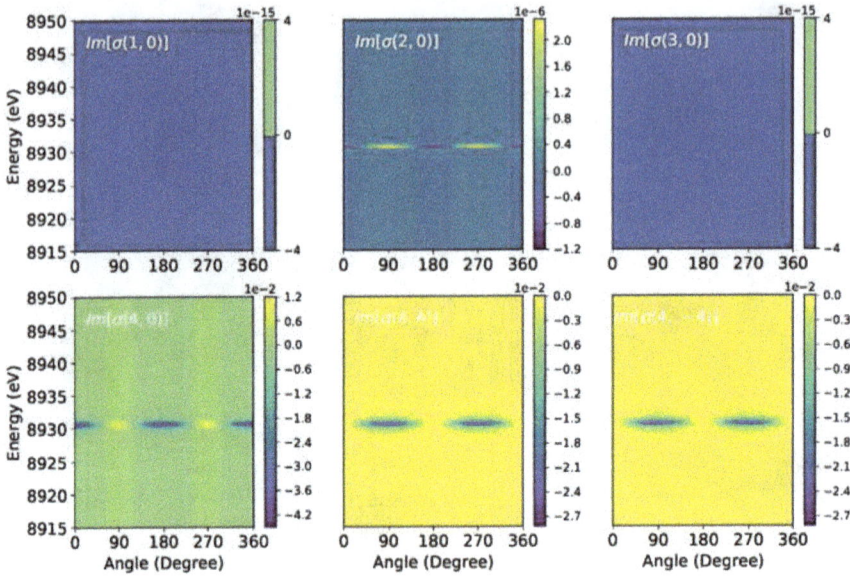

Fig. 4.19 Angular dependence of XAS fundamental cross sections $\sigma(1, 0)$, $\sigma(2, 0)$, $\sigma(3, 0)$, $\sigma(4, 0)$, $\sigma(4, 4)$, and $\sigma(4, -4)$. Here k is aligned with [100] and ϵ is rotated about [100] with $\theta = 0°$ for $\epsilon \parallel$ [001]

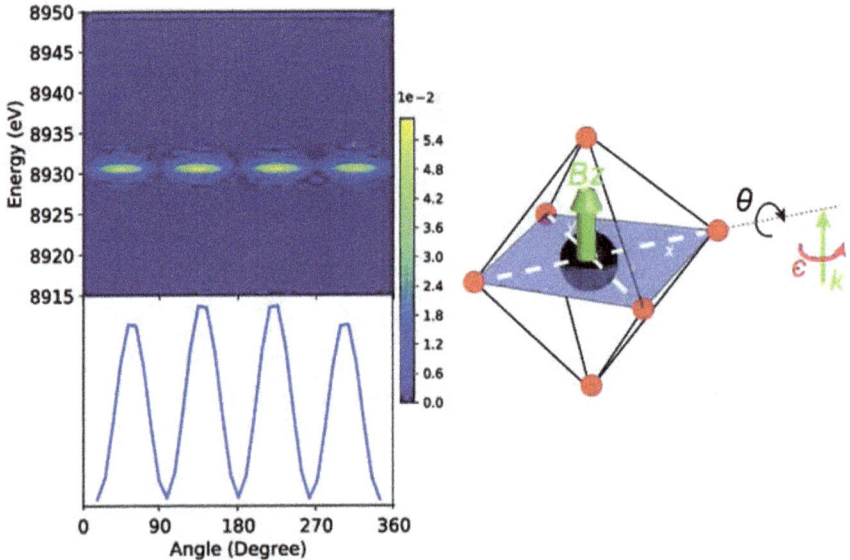

Fig. 4.20 Angular dependence of XAS for a quadrupole transition in a $3d^9$ ion in an O_h crystal field ($10D_q = 1.1$ eV) with an exchange field aligned along the z-axis ($B = 0.05$ eV). The wave vector (k) is initially aligned to [001] and polarization (ϵ) is circular. The angular dependence is computed by rotating the system about the [100] axis

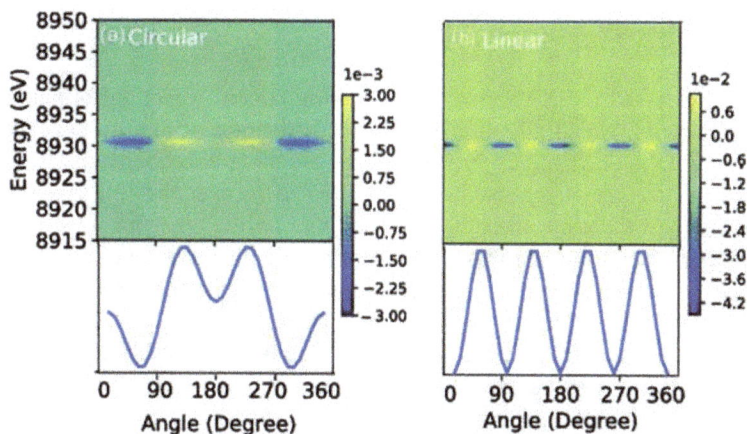

Fig. 4.21 Angular dependence of XAS fundamental spectra for a quadrupole transition in a $3d^9$ ion in an O_h crystal field with an exchange field aligned along the z-axis. **a** Circular dichroism active terms $\sigma(1, 0)$ and $\sigma(3, 0)$. **b** Linear dichroism active terms $\sigma(4, 0)$, $\sigma(4, 4)$ and $\sigma(4, -4)$. The angular dependence is computed by rotating the system about the [100] axis

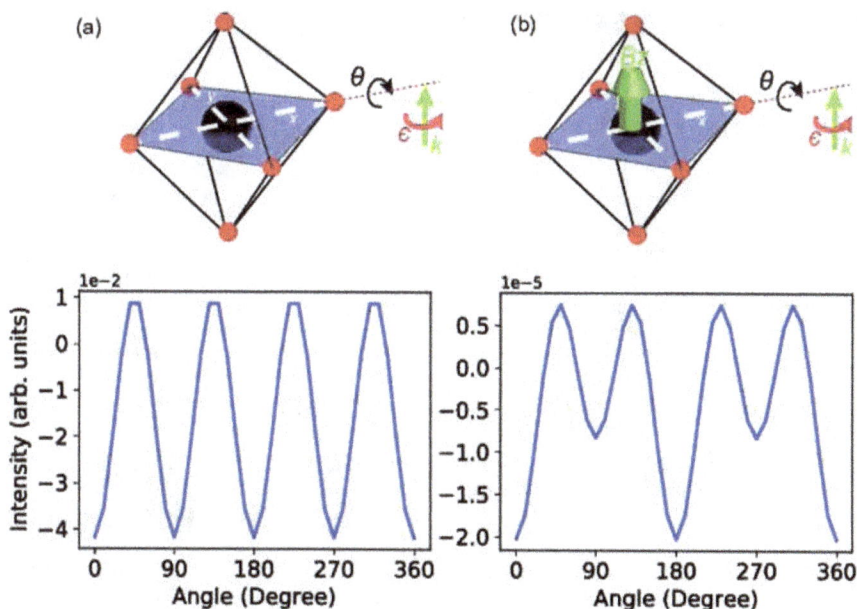

Fig. 4.22 Angular dependence of XAS for a quadrupole transition in a $3d^9$ ion: **a** in an O_h crystal field ($10D_q = 1.1$ eV). **b** The difference between calculation (**a**) and in an O_h crystal field with an exchange field aligned along the z-axis ($10D_q = 1.1$ eV and $B = 0.05$ eV). The wave vector ($\boldsymbol{k}$) is initially aligned to [001] and polarization ($\boldsymbol{\epsilon}$) is circularly polarized. The angular dependence is computed by rotating the system about the [100] axis

4.3   Conclusion

We have expressed the XAS cross section using Green's function formalism and spherical tensors, which allows a tractable investigation of the different types of dichroism: on the one hand experimentally, by allowing the experimentalist to predict the smallest set of measurements required to recover the full spectroscopic information and therefore to optimize beamtime; on the other hand theoretically, using modern core level spectroscopy codes, by allowing a more efficient calculation rather than a point-by-point time-consuming treatment. In a forthcoming work, we intend to use a similar approach for resonant inelastic X-ray scattering (RIXS) spectroscopy, whose richness lies in the large number of possible spectra that can be obtained by varying the energy, direction, and polarization state of the incident and scattered beams [28]. But as a matter of fact, there is so much information in the spectra that it is difficult to know whether a specific set of experiments measures all potential information. In our opinion, the possibilities offered by angular and polarization dependent RIXS measurements have not yet been exploited to the best of their potential. Nevertheless, the technique has now become mature and popular and will benefit from the huge instrumentation progress achieved in the last years, which open new doors for the exploration of dichroims in materials.

References

1. C. Brouder, Angular dependence of X-ray absorption spectra. J. Phys.: Condens. Matter **2**, 701 (1990). https://doi.org/10.1088/0953-8984/2/3/018
2. A. Juhin, S.P. Collins, Y. Joly, M. Diaz-Lopez, K. Kvashnina, P. Glatzel, C. Brouder, F.M.F. de Groot, Measurement of f orbital hybridization in rare earths through electric dipole-octupole interference in X-ray absorption spectroscopy. Phys. Rev. Mater. **3**, 120801(R) (2019). https://doi.org/10.1103/PhysRevMaterials.3.120801
3. S. Sugano, T. Tanabe, H. Kamimura, *Multiplets of Transition Metal Ions in Crystals* (Academic, New York, 1970)
4. J.S. Griffith, *The Theory of Transition-Metal Ions* (Cambridge University Press, Cambridge, 1961)
5. E. König, S. Kremer, *Ligand Field Energy Diagrams* (Springer Science+Business Media, New York, 2013). https://doi.org/10.1007/978-1-4757-1529-3
6. R.D. Cowan, *The Theory of Atomic Structure and Spectra* (University of California Press, Berkeley, 1981)
7. M.W. Haverkort, Y. Lu, S. Macke, R. Green, M. Brass, S. Heinze, Quanty – a quantum many body script language, http://www.quanty.org/doku.php?id=start&rev=1522759229
8. M. Baer, Multipole expansion package, https://github.com/maroba/multipoles
9. M.W. Haverkort, M. Zwierzycki, O.K. Andersen, Multiplet ligand-field theory using Wannier orbitals. Phys. Rev. B **85**, 165113 (2012). https://doi.org/10.1103/PhysRevB.85.165113
10. Y. Lu, M. Höppner, O. Gunnarsson, M.W. Haverkort, Efficient real-frequency solver for dynamical mean-field theory. Phys. Rev. B **90**, 085102 (2014). https://doi.org/10.1103/PhysRevB.90.085102
11. M.W. Haverkort, G. Sangiovanni, P. Hansmann, A. Toschi, Y. Lu, S. Macke, Bands, resonances, edge singularities and excitons in core level spectroscopy investigated within the dynamical

mean-field theory. Europhys. Lett. **108**, 57004 (2014). https://doi.org/10.1209/0295-5075/108/57004

12. M.W. Haverkort, Quanty for core level spectroscopy - excitons, resonances and band excitations in time and frequency domain. J. Phys. Conf. Ser. **712**, 012001 (2016). https://doi.org/10.1088/1742-6596/712/1/012001

13. M. Retegan, Crispy: v0.7.3. https://doi.org/10.5281/zenodo.1008184

14. C. Brouder, A. Juhin, A. Bordage, M.-A. Arrio, Site symmetry and crystal symmetry: a spherical tensor analysis. J. Phys.: Condens. Matter **20**, 455205 (2008). https://doi.org/10.1088/0953-8984/20/45/455205

15. B.T. Thole, G. van der Laan, Spin polarization and magnetic dichroism in photoemission from core and valence states in localized magnetic systems. Phys. Rev. B **44**, 12424 (1991). https://doi.org/10.1103/PhysRevB.44.12424

16. G. van der Laan, B.T. Thole, Spin polarization and magnetic dichroism in photoemission from core and valence states in localized magnetic systems. II. Emission from open shells. Phys. Rev. B **48**, 210 (1993). https://doi.org/10.1103/PhysRevB.48.210

17. B.T. Thole, G. van der Laan, Spin polarization and magnetic dichroism in photoemission from core and valence states in localized magnetic systems. III. Angular distributions. Phys. Rev. B **49**, 9613 (1994). https://doi.org/10.1103/PhysRevB.49.9613

18. G. van der Laan, B.T. Thole, Spin polarization and magnetic dichroism in photoemission from core and valence states in localized magnetic systems. IV. Core-hole polarization in resonant photoemission. Phys. Rev. B **52**, 15355 (1995). https://doi.org/10.1103/PhysRevB.52.15355

19. G. van der Laan, B.T. Thole, Core hole polarization in resonant photoemission. J. Phys. Condens. Matter **7**, 9947 (1995). https://doi.org/10.1088/0953-8984/7/50/028

20. P. Carra, H. König, B.T. Thole, M. Altarelli, Magnetic X-ray dichroism: general features of dipolar and quadrupolar spectra. Physica B **192**, 182 (1993). https://doi.org/10.1016/0921-4526(93)90119-Q

21. J.-P. Schillé J.-P. Kappler, P. Sainctavit, C. Cartier dit Moulin, C. Brouder, G. Krill, Experimental and calculated magnetic dichroism in the Ho $3d$ X-ray absorption of intermetallic $HoCo_2$. Phys. Rev. B **48**, 9491 (1993). https://doi.org/10.1103/PhysRevB.48.9491

22. B.T. Thole, G. van der Laan, M. Fabrizio, Magnetic ground-state properties and spectral distributions. I. X-ray-absorption spectra. Phys. Rev. B **50**, 11466 (1994). https://doi.org/10.1103/PhysRevB.50.11466

23. A. Juhin, C. Brouder, M.-A. Arrio, D. Cabaret, P. Sainctavit, E. Balan, A. Bordage, A.P. Seitsonen, G. Calas, S.G. Eeckhout, P. Glatzel, X-ray linear dichroism in cubic compounds: the case of Cr^{3+} in $MgAl_2O_4$. Phys. Rev. B **78**, 195103 (2008). https://doi.org/10.1103/PhysRevB.78.195103

24. G. van der Laan, Magnetic linear X-ray dichroism as a probe of the magnetocrystalline anisotropy. Phys. Rev. Lett. **82**, 640 (1999). https://doi.org/10.1103/PhysRevLett.82.640

25. E. Arenholz, G. van der Laan, R.V. Chopdekar, Y. Suzuki, Anisotropic X-ray magnetic linear dichroism at the Fe $L_{2,3}$ edges in Fe_3O_4. Phys. Rev. B **74**, 094407 (2006). https://doi.org/10.1103/PhysRevB.74.094407

26. G. Schütz, W. Wagner, W. Wilhelm, P. Kienle, R. Zeller, R. Frahm, G. Materlik, Absorption of circularly polarized X rays in iron. Phys. Rev. Lett. **58**, 737 (1987). https://doi.org/10.1103/PhysRevLett.58.737

27. P. Carra, B.T. Thole, M. Altarelli, X. Wang, X-ray circular dichroism and local magnetic fields. Phys. Rev. Lett. **70**, 694 (1993). https://doi.org/10.1103/PhysRevLett.70.694

28. A. Juhin, C. Brouder, F. de Groot, Angular dependence of resonant inelastic X-ray scattering: a spherical tensor expansion. Cent. Eur. J. Phys. **12**, 323 (2014). https://doi.org/10.2478/s11534-014-0450-2

5

Synchrotron Radiation Based Measurements and Spintronic Systems

Richard Mattana, Nicolas Locatelli and Vincent Cros

Abstract Having access to the electronic and magnetic properties of spintronic systems is of crucial importance in view of their future technological developments. Our purpose in this chapter is to elaborate how a variety of synchrotron radiation-based measurements provides powerful and often unique techniques to probe them. We first introduce general concepts in spintronics and present some of the important scientific advances achieved in the last 30 years. Then we will describe some of the key investigations using synchrotron radiation concerning voltage control of magnetism, spin-charge conversion and current-driven magnetization dynamics.

5.1 General Introduction to Spintronics: From Magnetoresistive Effects to the Physics of Spin-Transfer Phenomena

Whereas electronics relies on the charge of carriers (electrons and holes) to create and convey information, in spintronics the spin of these carriers is also used. The generation of spin currents, their detection as well as their manipulation are the basic principles of any spintronic devices. Since the discovery of giant magnetoresistance (GMR) effect in 1988, several breakthroughs have been achieved and strongly impacted several domains of applications such as, e.g. magnetic sensors, hard-disk drive read heads and more recently nonvolatile memories to cite the most emblematic ones. In this first part, we aim at introducing the basic principles of GMR and some of the pioneer experiments. Then we present some other important breakthroughs such as tunnelling magnetoresistive effects and the manipulation of magnetization without the application of a magnetic field but using spin-transfer effects.

R. Mattana (✉) · N. Locatelli · V. Cros
Unité Mixte de Physique, CNRS, Thales, Université Paris-Saclay, 1 avenue Augustin Fresnel, 91767 Palaiseau, France
e-mail: richard.mattana@cnrs-thales.fr

5.1.1 Giant Magnetoresistance: An Historical Point of View

5.1.1.1 Electric Conduction in Ferromagnets

In $3d$ ferromagnetic (FM) metals, electric conduction comes mainly from $4s$ band electrons, whereas magnetism originates from electrons in $3d$ bands. Resistivity in such materials arises from scattering events of carriers on impurities, phonons and local potentials. In particular, localized d orbitals act as diffusive centres for the s electrons which carry most of the charge current. Moreover, the non-zero net local magnetization comes up with a shift of the d-bands for the two electron spins (up and down), leading to different densities of states (DOS) at the Fermi level (Fig. 5.1 left). At low temperature, the electron spin is observed to be conserved during different sources of scattering (electron–phonon, electron–electron, electron–magnon, ...), allowing to describe the conduction properties in a ferromagnet using a "two-current" model [1]. Spin-up and spin-down electrons carry the current in two separated channels (Fig. 5.1 right). The difference in DOS of $3d$ electrons for spin-up and spin-down channels then leads to different number of scattering events for the two sub-bands, and thus different resistivities. Conduction in ferromagnet can hence be characterized by the spin-asymmetry coefficient α defined by the ratio between the resistivity of spin-down channel over the resistivity of spin-up channel. The spin-asymmetry coefficient is then strongly correlated to the asymmetry of the DOS at the Fermi level. For example, Fe has a spin-asymmetry coefficient lower than 1 whereas for Co and Ni this coefficient is larger than 1.

The experimental demonstration of the two-current model has been provided by a series of experiments in which Fert and Campbell showed that the conduction of a spin channel can be tuned by introducing different types of magnetic impurity [2–5]. Depending on the introduction of two different impurities with similar or opposite spin asymmetries, the impact on the two spin channels will be different (Fig. 5.2). Let us take a Ni matrix ($\alpha > 1$) doped with Co impurities ($\alpha > 1$) and Rh impurities ($\alpha < 1$). With the presence of only Co impurities, scattering in

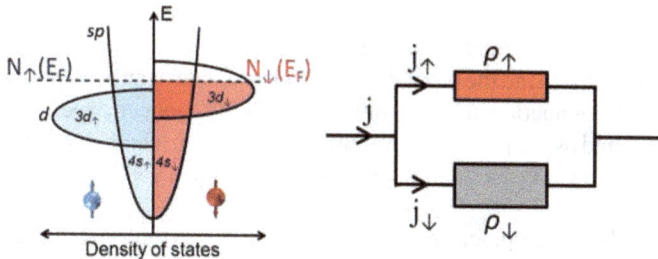

Fig. 5.1 (left) Scheme of the density of states in a ferromagnetic metal which has d bands shifted. (right) Scheme illustrating the two-current model: spin-up and spin-down electrons carry the current in separated channels. As resistivity arises from s-d transition and is proportional to the $3d$ DOS, the resistivity of a ferromagnet can be schematized by two resistances ($\rho_\uparrow$ and $\rho_\downarrow$) in parallel

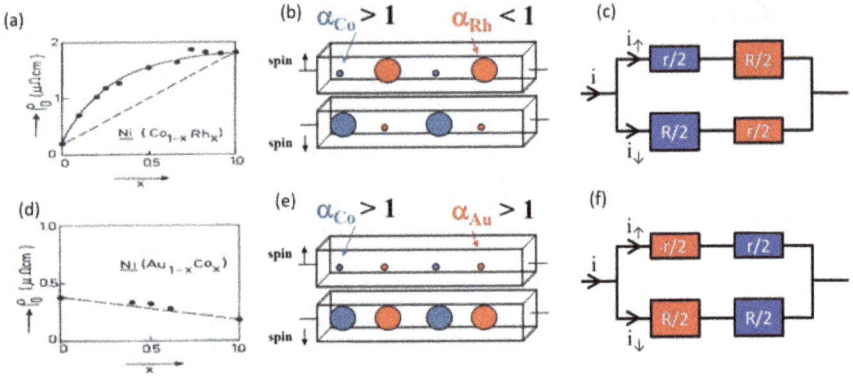

Fig. 5.2 Tailoring the channel conduction with impurities. **a** Resistivity of Ni as a function of Co and Rh doping. Co and Rh having opposite spin asymmetries, scattering is significant for both spin-up and spin-down channels **b** and **c** leading to an increase of resistivity. The resistivity dependence as a function of Co and Rh concentration is not linear **a**. **d** Resistivity of Ni as a function of Co and Au doping. Co and Au having similar spin asymmetries, electrons of only spin-down channel are strongly scattered **e**. The spin-up channel is providing a short-circuit for the electrons **f**. A small linear dependence of the resistivity with Co and Au impurity concentrations is then obtained **d**.

the spin-down channel is enhanced, but the spin-up channel still experiences low resistivity: the resulting global resistivity is low (short-circuit effect). As Rh impurities are introduced, scattering becomes significant for both channels [Fig. 5.2(b,c)], progressively suppressing the short-circuit effect, and the global resistivity quickly increases. The resistivity dependence as a function of Co and Rh concentrations is not linear [Fig. 5.2a]. Instead, when Rh impurities are replaced by Au ones ($\alpha > 1$, i.e. more electrons of the down channel are scattered, like for Ni and Co), such an enhancement of resistivity is not observed. Only electrons of the spin-down channel are strongly scattered by Co and Au impurities and a small linear dependence of the resistivity with Co and Au impurity concentrations is observed [Fig. 5.2d]. The spin-up channel is providing a short-circuit for the electrons [Fig. 5.2e, f].

These experiments indeed brought experimental demonstration of the validity of the two-current model and the proof that the mean resistivity can be varied by tailoring of the channel conduction through impurities. These pioneer experiments can be indeed considered as the pre-concept of the GMR effect in which different resistance states will be achievable by applying a magnetic field in a single heterostructure.

Thanks to the development of molecular beam epitaxy of metallic materials in the late 80's, it became hence possible to grow nanometre-thick magnetic metallic heterostructures (multilayers) involving several separated FM layers. Controlling the magnetic configurations of the magnetic films (parallel or antiparallel configurations of their magnetizations) then grants the possibility to tailor the spin-dependent transport of the system. This has led to the discovery of the GMR effect in 1988.

5.1.1.2   First Observation of GMR Effects

The first studies of GMR effects were done in Fe/Cr superlattices (Fe: FM, Cr: non-FM metal). A key point of these heterostructures was the control of the magnetic configuration. Exploiting the indirect exchange interaction between the magnetic layers transmitted by the conduction electron (called RKKY interaction) through the non-magnetic spacer layer, P. Grünberg and co-workers proved in 1986 that it is possible to align the magnetization of consecutive Fe layers in opposite direction at zero magnetic field by choosing adequate Cr interlayer thickness [6]. The parallel magnetic configuration can then be reached by applying a large magnetic field in order to overcome the RKKY coupling. First magnetoresistance curves have been obtained by the groups of A. Fert in Orsay, France (Fig. 5.3 left) and P. Grünberg in Jüllich, Germany (Fig. 5.3 right). A higher resistance state is observed at zero magnetic field at which the antiparallel magnetic configuration is stabilized. By applying a magnetic field, the parallel magnetic configuration is progressively reached inducing a large decrease of the resistance. In the experiments done in Grünberg's group, a GMR effect of 1.5% at low temperature was obtained in the case of a Fe/Cr/Fe trilayer (Fig. 5.3 right). The striking point is that this effect is about one order of magnitude larger in amplitude than the anisotropic magnetoresistance of a single Fe thin film (also shown on Fig. 5.3 right). In Fert's group, measurements were performed in Fe/Cr superlattices instead of trilayers, which allowed to considerably increase the amplitude of the effect that reached 80% for sixty Fe/Cr bilayers (Fig. 5.3 left).

An important development towards the implementation of the GMR systems for applications in read heads for hard-disk drives or sensors was made in 1991 by B. Dieny et al. [7] at IBM who reported the first observation of GMR in simple

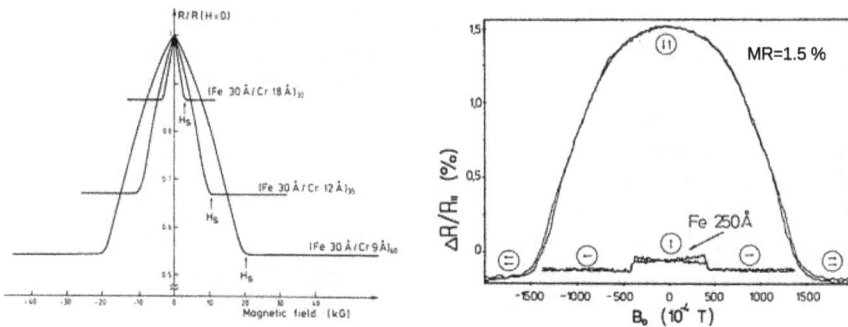

Fig. 5.3 Pioneer observation of GMR in Fe/Cr superlattices (left) and in Fe/Cr/Fe trilayer (right). At zero magnetic field, magnetization of consecutive layers is pointing in opposite direction leading to a high resistance state. By applying a magnetic field, a parallel magnetic configuration is reached and so a low resistance state is obtained. For a Fe/Cr/Fe trilayer, the GMR ratio is about 1.5% and one order of magnitude larger than the anisotropic magnetoresistance of a single Fe layer (right). By using Fe/Cr superlattices (60 Fe/Cr repetitions) the GMR ratio reaches 80% at low temperature.

trilayered structures with two distinct coercive fields, later called spin valves. In these structures, one of the layers switches using low fields while the second is stable up to large fields. It is to be emphasized that the time between the first discovery in a lab and the use of the GMR effect in cutting-edge technology devices has been extremely fast as less than 10 years after the discovery of GMR, and IBM introduced in 1997 spin valves in read heads of hard-disk drives. GMR-based magnetic sensors is now used in a multitude of applications such as monitoring wheel speed, detecting charge current, and fluid flow.

The research on magnetic multilayers and GMR became rapidly a very hot topic. It is not our purpose here to make here a review of all experimental and theoretical results that followed up the pioneer results. A complete review can be consulted in [8].

5.1.1.3   A Simple Model to Describe the GMR

The *two-current model*, described previously, was developed to explain the spin-dependent resistivity in magnetic materials. It can, in a rather simple way, be adapted to describe the giant magnetoresistive effect in these magnetic multilayers. This model is based on two assumptions: 1) $\alpha > 1$: the minority-spin electrons (opposed to local magnetization) are more scattered than those of majority spin (aligned with local magnetization); 2) the spin is conserved when the electrons are scattered. These two conditions are fulfilled at low temperature.

Two geometries can be considered to evaluate the resistance of such a structure: either the current flows in the direction of the layer planes (known as "current-in-plane GMR", CIP-GMR), or the current flows in a direction perpendicular to the layer plane (known as "current-perpendicular-to-the-plane GMR", CPP-GMR). A similar description can be used to account for the magnetoresistive properties for both geometries, without entering into the fine discussion about the actual physical spin-transport mechanisms, as long as the layer thickness remains small compared to a characteristic length associated with each geometry: the mean free path for the CIP case and the spin-flip length for the CCP case [11].

Let us note r the resistance of a FM layer for the majority-spin channel and R the resistance for the minority-spin channel, with $r < R$. Here the resistance of the NM separating layer is neglected. As illustrated in Fig. 5.4, an electron will have to pass through adjacent FM layers. Depending on whether these layers have parallel (P) or anti-parallel (AP) magnetizations, the resulting resistance shall differ.

In the (P) configuration, the electrons with spin ($\uparrow$) and ($\downarrow$) are either the electron spin majority or minority in all magnetic layers. Spin ($\uparrow$) electrons then experience identical resistances $r_\uparrow = r$ when crossing each magnetic layer, while spin ($\downarrow$) electrons experience identical resistances $r_\downarrow = R$. The mean resistance then writes $r_P = \frac{Rr}{r+R}$. In case of materials with large spin asymmetry ($\alpha \gg 1$ and $r \ll R$), the multilayer is short-biased by the spin ($\uparrow$) channel and the total resistance is $r_P \approx r$.

For the (AP) configuration, the electrons of the two ($\uparrow$) and ($\downarrow$) channels correspond alternately to electron spin majority and minority in consecutive magnetic

Parallel configuration (P)   **Anti-parallel configuration (AP)**

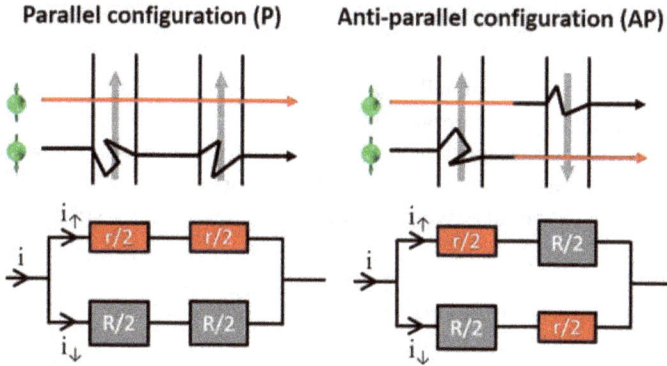

Fig. 5.4 Illustration of the two-current model: conduction of an electron in a ferromagnetic metal/normal metal/ferromagnetic metal (FM/NM/FM) multilayer. The electrons with a spin aligned parallel (resp. anti-parallel) to the local magnetization see a resistance r (resp. R) through this magnetic layer. The equivalent resistance circuit is presented for two configurations: consecutive ferromagnetic layers with parallel magnetization or anti-parallel magnetization

layers, and thus the effect of short-circuit by one of the channels is suppressed. Consequently the two channels have the same resistance $\frac{R+r}{2}$ and the total resistance is $\frac{R+r}{4}$, which is in general much larger than $r_P = r$.

This brings the possibility of switching between high and low resistance states by simply changing the relative orientation of the magnetization of consecutive layers. If we can engineer these structures to stabilize the two (P) and (AP) magnetic states in absence of any external field, then such system defines a magnetic bit to store the information, using the powerful stability of the magnets, already well known through the development of hard-disk drives. The state can then be read through a very simple process, i.e. by measuring the resistance of the stack.

Following this model, the amplitude of the GMR ratio can be simply deduced

$$GMR = \frac{R_{AP} - R_P}{R_P} = \frac{(R-r)^2}{4Rr} \tag{5.1}$$

In short, the understanding and expertise acquired during the 60s and 70s for tailoring the spin-transport properties combined with the development of metallic magnetic multilayers in the 80s has led to the discovery of the GMR effect in 1988. Less than 10 years after the discovery of the GMR effect considered as the birth of spintronics, this effect has largely participated to the explosion of the amount of data storage in the mid-90s through the commercialization of the first generation of spintronic devices such as magnetic sensors used as read heads in hard-disk drives.

5.1.2   Tunnelling Magnetoresistance

The pioneer observation of tunnelling magnetoresistance (TMR) in magnetic tunnel junctions (MTJs) at room temperature in 1995 is considered to be the second breakthrough in spintronics, leading to a second generation of spintronic applications such as magnetic random access memory (MRAM). In this section, we describe the first TMR measurements, then the two standard models describing TMR effects.

5.1.2.1   First Experimental and Theoretical Studies

A magnetic tunnel junction is a device in which two FM electrodes are separated by an ultra-thin insulating barrier. The first MTJ exhibiting TMR has indeed been reported in 1975 by M. Jullière [12] who measured a TMR ratio of 14% at 4.2K in a Fe/Ge/Co junction. Twenty years later TMR effects have been observed at room temperature by J. S. Moodera [13] and T. Miyazaki [14] using amorphous alumina as tunnel barrier. In Fig. 5.5, the TMR curve obtained for a CoFe/Al$_2$O$_3$/Co MTJ [13] is displayed. The TMR ratio is around 12%, at room temperature. This increase in the resistance difference between the two states allows an easier detection of them and let appear the potential of such spintronic devices for memory applications.

Such TMR effect has been first explained by M. Jullière in 1975. Keeping the free-electron approximation, he proposed an additional assumption: the *electron spin conservation* during the tunnelling process, meaning that electrons can tunnel from one FM electrode toward the second FM electrode only into empty states having identical spins. This simple model easily allows to understand that the tunnelling current in the (P) and (AP) alignments of magnetizations will differ.

As illustrated in Fig. 5.5, for the parallel (P) configuration of the magnetizations: when a small voltage is applied, majority-spin (denoted as ↑) electrons from the injector will tunnel toward majority-spin (↑) empty states of the collector, allowing "high" current. At the same time, minority-spin electrons from the injector will tunnel toward minority-spin (↓) empty states of the collector, allowing "low" current.[1] The conductance then writes

$$G_P \propto D_1^{\uparrow}(E_F)D_2^{\uparrow}(E_F) + D_1^{\downarrow}(E_F)D_2^{\downarrow}(E_F) \; . \tag{5.2}$$

In the case of antiparallel (AP) configuration of the magnetizations, majority-spin (↑) electrons from the injector will tunnel toward minority-spin (↓) empty states of the collector. Minority-spin (↓) electrons from the injector will tunnel toward majority-spin (↑) empty states of the collector. Both currents are then low. The conductance then writes

---

[1] Note here that the arrow does not define any direction of the spin, but rather the fact that it is aligned with the local magnetization. In the (AP) case, a spin with one direction will correspond to (↑) spin in one electrode and (↓) spin in the other.

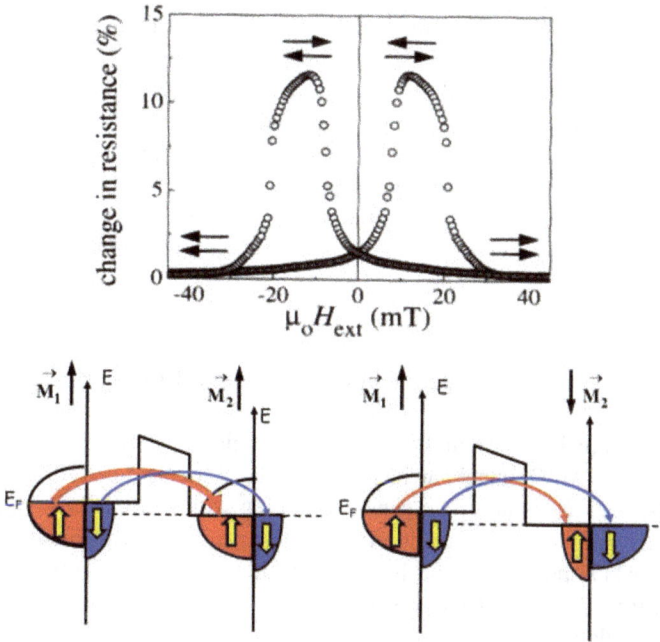

Fig. 5.5 TMR curve of a CoFe/Al$_2$O$_3$/Co MTJ recorded at room temperature [Adapted from [13] with permission]. Schematic of the spin-dependent tunnelling process through an insulating barrier when their magnetizations are aligned parallel (left) and antiparallel (right) to one another. The process is assumed to be purely elastic, so that no mixing of spin states occurs during the tunnelling process

$$G_{AP} \propto D_1^{\uparrow}(E_F)D_2^{\downarrow}(E_F) + D_1^{\downarrow}(E_F)D_2^{\uparrow}(E_F) \ . \qquad (5.3)$$

Using these results for the conductance, the TMR ratio, characterizing the resistance difference in (P) and (AP) configurations, is expressed as

$$TMR = \frac{G_P - G_{AP}}{G_{AP}} = \frac{R_{AP} - R_P}{R_P} = \frac{2P_1 P_2}{1 - P_1 P_2} \ , \qquad (5.4)$$

where P_i defines the spin polarization for each electrode as

$$P_i = \frac{D_i^{\uparrow}(E_F) - D_i^{\downarrow}(E_F)}{D_i^{\uparrow}(E_F) + D_i^{\downarrow}(E_F)} \ . \qquad (5.5)$$

A key issue to get an accurate prediction of TMR is to properly estimate the actual amplitude and sign of the spin polarization for a given FM material. Obviously, the largest the spin polarization is, the highest will be the TMR amplitude. It thus explains the strong research activity in the last decade on material science to investigate novel families of materials (magnetic oxides, Heusler alloys, etc.) for which some of their compounds are predicted to be half metallic, i.e. to have a 100% spin polarization at the Fermi level.

Jullière's model has enjoyed much success in correctly predicting TMR amplitudes in most MTJs with amorphous barriers, until the emergence of MTJs based on crystalline insulating barriers, notably magnesium oxide (MgO) barriers in the beginning of the 2000s [15, 16]. The model described previously is actually oversimplified in that it considers an identical tunnelling process of every electron, independently on their band belonging, hence forgetting about the interplay of the band structures throughout the MTJ heterostructure.

5.1.2.2   Coherent Tunnelling in Epitaxial Magnetic Tunnel Junctions

Another major breakthrough in the field of spintronics corresponds to the introduction of crystalline MgO tunnel barriers, that is today's standard in MTJ implemented in MRAMs, read heads and field sensors working with TMR effect. These experimental developments have been stimulated by some theoretical calculations made by W. Butler [17] predicting that huge TMR ratios as large as 1600% are anticipated for *epitaxially grown* Co or Fe electrodes on crystalline, instead of amorphous MgO. These calculations, developed with *ab initio* methods, derive the tunnelling probability of each kind of electrons, depending on their orbital symmetry.

Contrary to amorphous alumina [Fig. 5.6a left], in crystalline MgO tunnel junctions, the electron wave functions in the FM material are coupled with evanescent wave functions having the same symmetry in the barrier [Fig. 5.6a right]. Through *ab initio* calculations, it was then predicted that the tunnelling probability of an electron strongly depends on the orbital symmetry of the electron (of the band it belongs to). Beyond the two-current (up-spin and down-spin) model, the system behaves as if it exists an independent current channel for each band and each spin, leading to a possible effective *symmetry filtering of the tunnelling current*. The tunnel barrier can, therefore, filter the wave functions and thus select the spins in the electronic transport.

This mechanism of orbital selection for the tunnel conductance is presented in Fig. 5.6. In Fig. 5.6b, band dispersion of *bcc* Fe(001) for the minority and majority spins is shown. Δ_1 Bloch states are present at the Fermi level only for the majority spins. In Fe(100)/MgO/Fe(100) systems, band structure calculations have demonstrated that majority-spin electrons [see Fig. 5.6c] are mainly filling Δ_1 symmetry states (hybridized states with *spd* characters), whereas minority-spin electrons [see Fig. 5.6d] are filling Δ_2 symmetry states (*d* type states) and Δ_5 symmetry states (hybridized *pd* states). Moreover, this is crucial for getting a large TMR ratio, the tunnelling exponential decay is much stronger for Δ_2 and Δ_5 states compared to Δ_1 states. For $d_{MgO} = 8$ monolayers, which is a typical barrier thickness that can be achieved experimentally, the probability of transmission of Δ_1 electrons is larger than for Δ_5 electrons by 10 orders of magnitude. Ultimately, only Δ_1 electrons contribute significantly to the current. It is this filtering effect which can explain the large values of TMR expected on epitaxial or highly textured structures and that has made

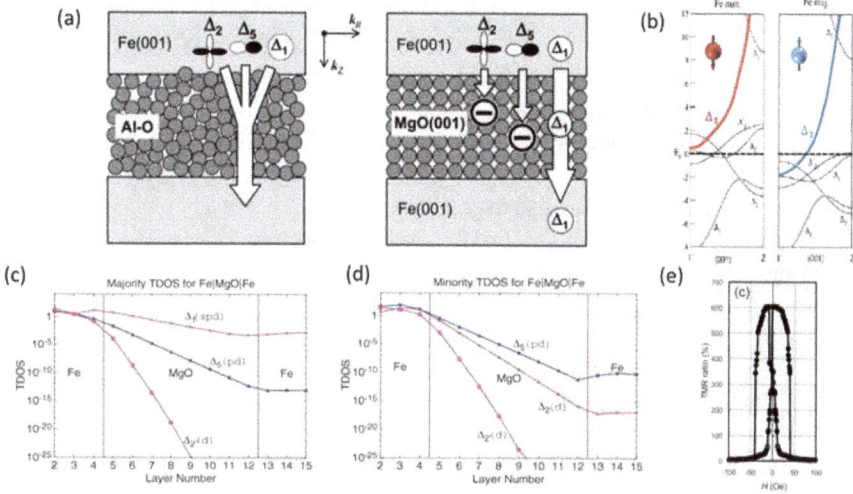

Fig. 5.6 **a** Schemes illustrating electron tunnelling through an amorphous Al-O barrier (left) and through a crystalline MgO barrier (right). **b** Band dispersion of *bcc* Fe(001) for the minority and majority spins. Δ_1 Bloch states are present at the Fermi level only for the majority spins. Tunnelling DOS for Fe/MgO/Fe at $k_\parallel = 0$ for majority **c** and minority **d** spins. Decay of Δ_1 Bloch states in MgO is less attenuated than the Δ_1 Bloch states. **e** TMR of 600% obtained at room temperature for a CoFeB/MgO/CoFeB MTJ.

the MgO-based tunnel junctions at the core of the development of new spintronic devices like the MRAMs.

After these theoretical predictions, a strong research effort has been made to obtain epitaxial growth of structures for Fe/MgO/Fe or CoFeB/MgO/CoFeB [20, 21]. These efforts have resulted in a TMR of about 600% obtained in 2009 [19] in CoFeB/MgO/CoFeB MTJs at room temperature [see Fig. 5.6e]. An excellent review on the physics of tunnelling transport in MgO-based systems and the experimental state of the art for TMR has been written by S. Yuasa [18].

Since the measurements of TMR effects at room temperature in 1995, a lot of work has been done in order to both increase the TMR ratio and reduce the MTJ resistance. This has been achieved by developing high-quality crystalline MgO-based MTJs. This effort has led to the development of a new class of magnetic memories called MRAMs.

5.1.3   Magnetization Manipulation without Magnetic Fields

For practical applications, e.g. MRAMs, the possibility to manipulate the magnetization of FM electrodes in spin valves or MTJs without using magnetic field is required. This corresponds to the next breakthrough with the prediction and the observation of spin-transfer effects, first providing a new way to reverse the magnetization in a nanostructure but also to generate in some cases steady precession of the magnetization. In this section, we briefly present the physics of spin-transfer torque (STT) and also discuss an alternative approach consisting in the manipulation of the magnetization through the application of a voltage.

5.1.3.1   Spin-Transfer Torque

In the first generation of MRAM devices developed in 2004, the magnetization reversal between the two possible magnetic states was realized by using local magnetic fields generated by electrical currents flowing in lines close to each magnetic element. This writing process rapidly suffered from both the large energy consumption needed to generate large enough magnetic fields and from cross-talk problems due to the difficulty to write a single bit. A solution came out from a major breakthrough in spintronics in 1996 when it has been proposed that a spin-polarized current flowing through magnetic multilayers provides a new way to manipulate the magnetization of a ferromagnet. This new effect, which was called spin-transfer effect, has become rapidly a very hot topic. Moreover, it is at the basis of a new generation of magnetic memories, called spin-transfer torque-MRAMs (STT-MRAMs).

The concept of spin-transfer effect was proposed in 1996, concomitantly by J. Slonczewski [22] and L. Berger [23]. To describe this effect, one can take a standard spintronic structure composed of a fixed FM electrode F_1 and a free FM layer F_2 separated by a NM spacer. When the electrons flow from layer F_1 to F_2, the current becomes spin-polarized after passing through F_1. This non-zero spin polarization is aligned along the magnetization direction in F_1 and propagates into the NM metallic spacer or tunnels in the case of an insulating barrier, so that it arrives at the NM/F_2 interface. If the magnetization $\overrightarrow{M_1}$ and $\overrightarrow{M_2}$ are non-collinear, it results that a component of the spin current transverse to F_2 exists at the NM/F_2 interface [green arrows in Fig. 5.7a]. When the electrons penetrate into F_2, the spin of the conduction electron becomes aligned over a very short distance (within a few atomic distances) toward the magnetization direction in F_2 because of a strong exchange interaction between conducting (s) electrons and localized (d) electrons, the latter being responsible of the magnetic moments.

During all this process, the total spin angular momentum is conserved. Thus the transverse component of the spin current $\Delta \overrightarrow{m}$ lost by the electrons when passing through F_2 is indeed absorbed and transferred to the local magnetization of F_2. This transfer of spin angular momentum results in a torque exerted by the spin-polarized current on the local magnetization. For this current polarity, when the

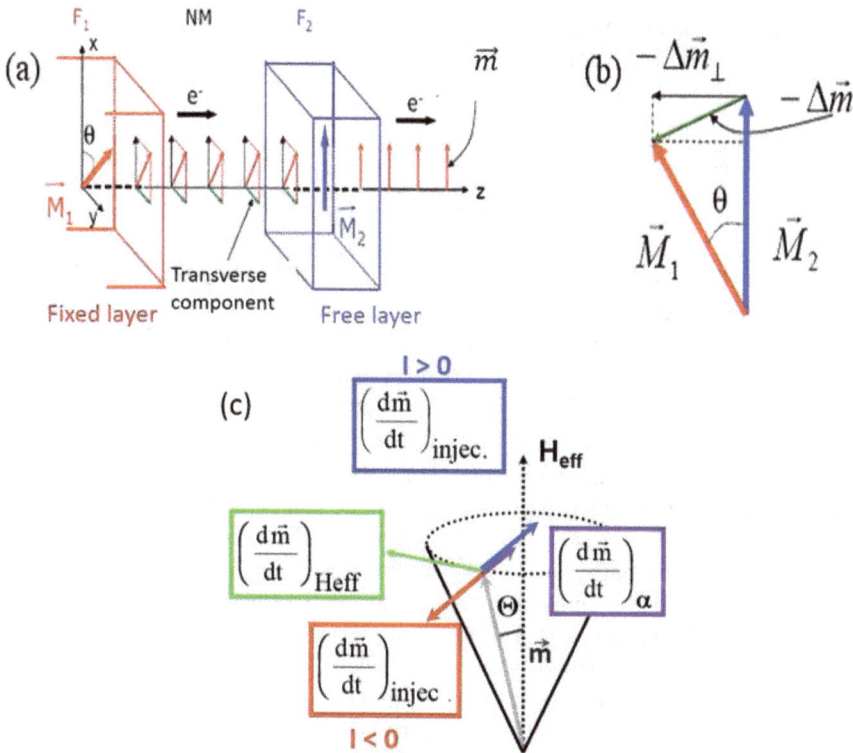

Fig. 5.7 **a** Scheme of a magnetic trilayer structure for illustrating the concept of spin-transfer torque. **b** The transverse component of the angular moment Δm of the spin-polarized current is transferred to the magnetization. It results in a torque exerted on the magnetization $\overrightarrow{M_2}$, of the thin magnetic layer, which aligns along magnetization $\overrightarrow{M_1}$ for positive current. **c** Schematic representation of the torques acting on the magnetization including spin-transfer effect

current increases enough, it tends to align the magnetization $\overrightarrow{M_2}$ along the direction of the spin polarization of the current, i.e. along the magnetization $\overrightarrow{M_1}$ [Fig. 5.7b]. As all the process occurs in the first atomic planes after the interface, the spin-transfer mechanism is an interfacial effect.

After having introduced the spin-dependent transport mechanisms at the origin of STT, our aim is now to address the influence of this torque on the magnetization dynamics of the layer F_2. A classical approach to describe the dynamical motion of a magnetization is a differential equation, named the Landau–Lifschitz–Gilbert equation, to which we add Slonczewski's component of spin torque.[2] This equation has the

---

[2]For simplicity we have only introduced the main STT called "in-plane torque". A second spin torque, called "field-like torque" or "out-of-plane torque", is similar to a torque exerted by a field along the magnetization of F_2 and its action might, therefore, be included into $\overrightarrow{H}_{eff}$.

following form, taking magnetizations normalized to the saturation magnetization, $m_{i=1,2}$ for the two FM layers

$$\frac{d\vec{m_2}}{dt} = -\gamma_0(\vec{m_2} \times \vec{H}_{eff}) + \frac{\alpha}{\gamma_0 M_{S2}}(\vec{m_2} \times \frac{d\vec{m_2}}{dt}) - P_{spin}\frac{J}{e}\frac{h}{2}\frac{g\mu_B}{M_{S2}^2 t}[\vec{m_2} \times (\vec{m_2} \times \vec{m_1})]$$

$$(5.6)$$

The first term corresponds to the tangential force describing the magnetization precession around effective field $\vec{H}_{eff}$ [green arrow in Fig. 5.7c], which takes into account the external applied magnetic field, magnetic anisotropy fields, coupling fields, etc. The second term gives a phenomenological description of magnetization dissipation of the system. The coefficient α, named Gilbert damping, (about 10^{-2} for standard FM materials), describes the damping rate of the motion of $\vec{m_2}$ towards the equilibrium position oriented along $\vec{H}_{eff}$ [blue arrow in Fig. 5.7c]. The magnetization relaxation is induced by a damping force tangential to the magnetization trajectory. The third term is the so-called Slonczewski torque (or in-plane torque) where μ_B is the Bohr magneton, t the layer thickness, J the injected current density, and P_{spin} the amplitude of spin polarization at the interface NM/F_2 and M_{S2} the magnetization of the ferromagnet F_2. This simplified description allows making clear the nature of the main contribution of the spin-transfer force that can be described as a non-conservative force acting in the same direction than the natural damping, i.e. perpendicularly to the magnetization trajectory. Depending on the sign of the injected current, the spin-transfer torque decreases or increases the effective damping [purple and red arrows in Fig. 5.7c]; it favours the stability of the parallel or the antiparallel magnetic configuration.

For large current density ($\sim 10^7$ A cm^{-2} for a typical material system), the STT fully compensates the damping torque, steady magnetization precession occurring at the ferromagnetic frequency (typical in the GHz range for the ferromagnetic materials and in the THz range for antiferromagnetic ones) can be established. The spin transfer induced magnetization dynamics can convert into oscillations of resistance through the magnetoresistive effect described previously, and in turn into a radiofrequency voltage signal. Spin-torque effect thus makes it possible to convert a *dc* current into a *rf* voltage and so to build microwave oscillators at the nanoscale.

For larger current density, the torque can become sufficient to commute the magnetization between the two stable configurations. A negative current will, for instance, destabilize the parallel magnetization configuration, while stabilizing the anti-parallel configuration, allowing to commute from P to AP configuration. Identically, a positive current allows to commute from AP to P configuration.

First experimental results allowing to confirm the theoretical predictions of the existence of STTs have been obtained by M. Tsoi et al., using point contact geometry for injection of a large current into a magnetic layer [24]. Then after, it has been demonstrated that magnetization commutation can be achieved back and forth by the STT effect in Co/Cu/Co spin valves [25]. Figure 5.8a represents magnetization rever-

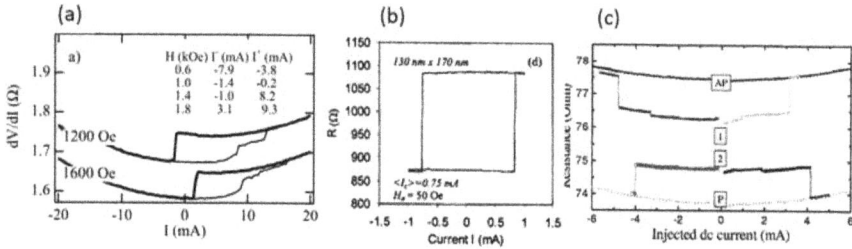

Fig. 5.8 a Current-induced magnetization switching in a Co/Cu/Co nanopillar spin valve. Jumps in the curves correspond to the magnetization reversal of the ferromagnetic free layer. **b** Current-induced magnetization switching in a CoFeB$_{2.5}$/AlOx/CoFeB$_{2.5}$ MTJ. **c** Domain wall motion induced by a current in Co/Cu/Py spin valves. State 1 and 2 correspond to different positions of the domain walls in the strip. By applying a large enough current it is possible to move the domain walls and thus reach the parallel (P) or antiparallel (AP) magnetic configuration.

sal induced by a current for two different applied magnetic fields. Current density in the 10^7 A cm^{-2} range corresponding to few mA for their nanopillars is needed to switch the Co free electrode magnetization. Few years later similar results have been obtained in MTJs using AlOx and MgO tunnel barriers [26]. Figure 5.8b illustrates the magnetization reversal by the current for a CoFeB/AlOx/CoFeB magnetic tunnel junction. A smaller current (less than 1 mA) is needed to commute the magnetization and thus to switch between the parallel and antiparallel magnetic configurations. This result shows that STT effects can be used not only in metallic spin valves but also in MTJs and has opened the door for the development of new STT-MRAM technologies. Finally, STT has been also used to induce magnetic domain wall motion. Figure 5.8c represents the first observation of domain wall motion by spin transfer in a Co/Cu/Py trilayer, where Py stands for permalloy (Ni$_{80}$Fe$_{20}$) [27]. Starting from configuration 1 or 2 (the domain wall is located at the two-third of the strip) it is possible to move the domain wall by a current to reach the parallel (P) or antiparallel (AP) magnetic configuration. The current density needed to move these domain walls is about 10^7 A cm^{-2}. Note that for all these pioneer experiments, a magnetic field is applied. Since then, a lot of work has been done and now magnetization manipulation without applying magnetic field is feasible. Later, STT has been also used to manipulate other magnetic textures such as magnetic vortices [28] or skyrmions, or generate dynamics [29, 30]. To conclude, STT is now used to write the information of a single bit in magnetic memories.[3]

---

[3]For more information on the latest MRAM development, see https://www.mram-info.com/

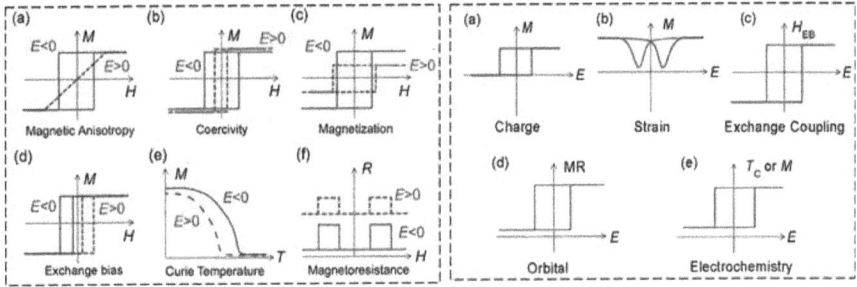

Fig. 5.9 (left panel) Voltage can be used to modify magnetic anisotropy, coercive fields, magnetization values, exchange bias, Curie temperature or magnetoresistance. This can be achieved by different mechanisms such as charge, strain exchange coupling, orbital ordering or electromigration (right panel).

5.1.3.2 Voltage Control of Magnetism

Although STT is very efficient to manipulate magnetization, alternative or complementary approaches have been explored such as the voltage control of magnetism. Magnetic anisotropy, coercive fields, magnetization magnitude, exchange bias or Curie temperature can be tuned by a voltage [Fig. 5.9 (left panel)]. Different mechanisms such as charge accumulation/depletion, electromigration, strain or orbital ordering are involved [Fig. 5.9 (right panel)].

The samples to study the effect of an electric field on the magnetization are generally composed of an ultra-thin FM film in contact with either dielectric, ferroelectric or piezoelectric materials. A wide variety of FM materials (metals, oxides or diluted magnetic semiconductors) have been also studied [31]. Depending on the mechanism used to control magnetism through an applied voltage, the ferromagnet thickness has to be adjusted. As charge, orbital ordering or electrochemistry are mainly interfacial effects, the thickness of the ferromagnet should be close to the screening length which is in the angstrom range for metal and in nanometre range for dilute magnetic semiconductors. Thicker ferromagnets, up to the micrometre range, can be used if strain is involved. In the following, several examples illustrating the variety of devices studied are displayed. Readers are invited to refer to the article of C. Song et al. [31] for a complete review.

In Fig. 5.10, we show different examples where the Curie temperature, the coercive field and the magnetic anisotropy are modified by an electric field. This magnetization control involves different mechanisms such as charge or strain effects. One of the first demonstration of magnetization manipulation by an electric field has been done using a dilute magnetic semiconductor [32]. In Fig. 5.10a, b, Hall effect curves recorded at different gate voltages for a (In,Mn)As-based field-effect transistor are presented. Magnetic properties and notably the Curie temperature of the diluted magnetic semiconductor (In,Mn)As being dependent on the hole density, it is thus possible to tune the Curie temperature by an electric field modulating the carrier density [33]. Setting the sample temperature close to its Curie temperature, the switching

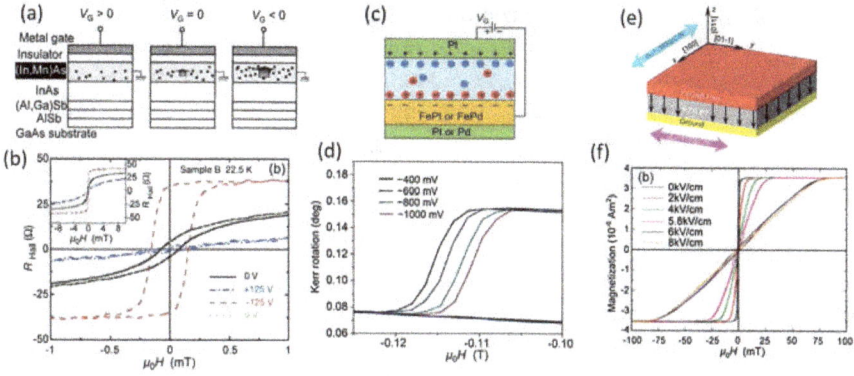

Fig. 5.10 Examples of voltage control of the magnetization. **a, b** Hall effect curves recorded for different gate voltages in a dilute magnetic semiconductor (In, Mn) As field-effect transistor. The gate voltage induces a modulation of the hole concentration in (In, Mn)As and thus a modulation of the Curie temperature. **c, d** Variation of the coercive field of a FePt thin film as a function of bias voltage. The bias voltage modifies the number of 3*d* electrons and thus changes the magnetocrystalline anisotropy. **e, f** Modification of the magnetic anisotropy induced by strain in a FeGaB/PZN-PT multi-ferroic heterostructure. The voltage leads to lattice modulation of the PZN-PT ferroelectric inducing a strain modulation of the magnetostrictive FeGaB layer. Under an electric field of $8\,kV\,cm^{-1}$ a magnetic field of 70 mT is needed to align the magnetization along the [100] direction.

between ferromagnetic and paramagnetic states can be realized by applying, respectively, a negative and a positive gate voltage (125 V). A similar approach has been investigated to tune the coercive field of the FePt FM intermetallic compound [34]. Using an electrolyte to modify electron density at the FePt interface, a modification of 4.5% of the coercive field is observed [Fig. 5.10c, d]. The variation in the number of 3*d* electrons directly affects the magneto-crystalline anisotropy and thus the coercive field. Interestingly, this device geometry allows tuning the magnetic properties of a ferromagnet by applying a low bias voltage (below 1 V). Although the modification of the coercitive film is quite small, this first demonstration of voltage control of a FM metal was quite encouraging as it opens the door to room temperature modulation of magnetism by applying small voltage. Another approach to modify the magnetic anisotropy is to apply a strain. By combining ferroelectric and magnetostrictive materials, it is possible to modify the magnetic properties by a voltage. Under voltage, the lattice of the ferroelectric layer is modulated through the inverse piezoelectric effect and will thus induce a strain modulation in the ferromagnet. In Fig. 5.10(e,f), we show an example where the in-plane magnetic anisotropy is modified by an electric field. Whereas at zero electric field, the remanent magnetization is along the [100] direction, an electric field of $8\,kV\,cm^{-1}$ allows rotating the magnetization and so a magnetic field of 70 mT is now needed to saturate it along the [100] direction.

The voltage control of magnetization is obviously less mature than STT or spin-orbit torque (SOT) technologies. However, the large variety of ferromagnets (metals, semiconductors, oxides) and gating materials (dielectric, ferroelectric, electrolyte) that can be used, makes this field fascinating and promising to reduce power consumption in memory technologies, which remains still a crucial issue.

5.1.4   Summary

We have introduced some breakthroughs in the field of spintronics achieved during the last 30 years. Giant magnetoresistance has emerged rapidly as a promising effect to build efficient magnetic sensors working at room temperature. The ability to manipulate magnetization by STT in MTJs has allowed developing new magnetic memories such as MRAMs that are now commercially available. This is of course not exhaustive and there are a lot of other exciting and promising research fields such as spin-orbitronics, magnonics, molecular spintronics or antiferromagnetic spintronics, to cite only few of them. Spin-orbitronics is certainly the most active field nowadays. Emblematic topics are conversion between charge and spin currents, spin-polarized surface and interface states or novel chiral magnetic textures (skyrmions and domain walls). See for example A. Soumyanarayanan et al. for a review [36].

In all the spintronic effects that we have just introduced, like in many other fields, interfaces play a key role, and therefore a deep knowledge of the electronic and magnetic properties of these interfaces is desired. Synchrotron radiation-based measurements such as absorption and photoelectron spectroscopies and microscopies are powerful techniques to probe these interfaces.

5.2   Examples of Synchrotron Radiation Contribution to Spintronics

In the following, several examples for which synchrotron radiation-based measurements have allowed a better understanding of spintronic devices are presented. We have selected three spintronics topics: (i) voltage control of magnetism; (ii) spintronics with pure spin currents; (iii) current-driven magnetization dynamics.

5.2.1 Voltage Control of Magnetism

5.2.1.1 Effect of Charge Accumulation/Depletion and Electromigration

Perpendicular magnetic anisotropy (PMA) has been widely investigated in NM/FM/oxide heterostructures. For example, by varying the oxidation time of top Al layer in Pt/Co/AlOx devices, it has been demonstrated that the PMA is induced by Co–O bonds [37]. Figure 5.11(a,b) represent extraordinary Hall effect curves and X-ray absorption spectroscopy (XAS) measurements at the Co $L_{2,3}$ edges for Pt/Co/AlOx samples with different oxidation times. The extraordinary Hall effect allows detecting the magnetic easy axis and XAS measurements probe the electronic properties of the cobalt layer. A metallic Co/Al interface resulting from Al under-oxidation, or the formation of a CoO layer due to over-oxidation, both produce magnetization aligned within the plane of the film. Hence optimized oxidation, leading to Co–O–Al bonds, induces a perpendicular magnetic anisotropy. Instead of playing with oxidation time, D. Chiba et al. [38] have shown that it was possible to switch the easy magnetic axis from in-plane to out-of-plane by applying a bias voltage. In Fig. 5.11c, we show that a bias voltage of 10 V allows to modify the magnetic

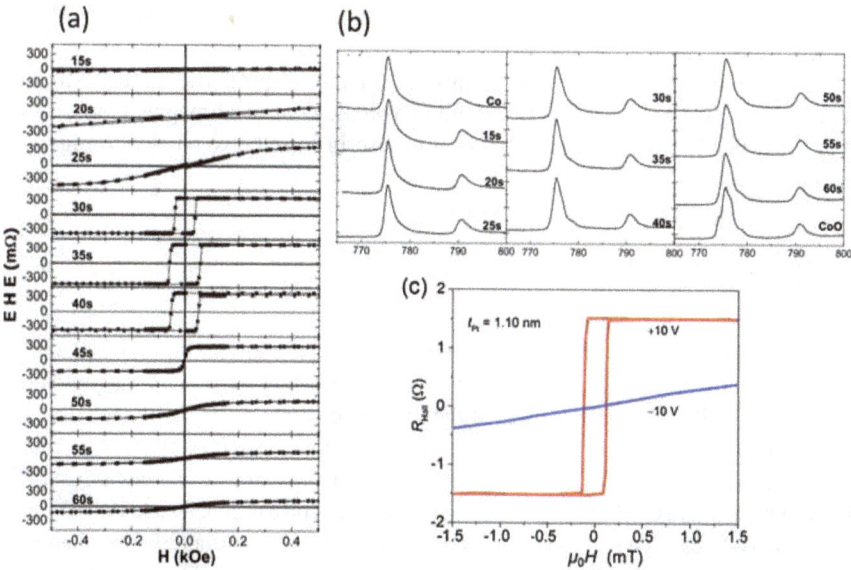

Fig. 5.11 a Extraordinary Hall effect as a function of oxidation time measured in a Pt/Co/AlOx trilayer. **b** XAS spectra recorded at the Co $L_{2,3}$ edges for different oxidation times. These measurements show that optimized Al oxidation induces perpendicular magnetic anisotropy due to Co–O bonds. **c** Switching between in-plane magnetization to out-of-plane magnetization by applying a bias voltage for a Pt/Co/MgO sample.

anisotropy in Pt/Co/MgO [38]. This magnetic anisotropy modification with voltage can originate from charge accumulation or oxygen electromigration.

C. Bi et al. have performed Hall resistivity and XAS measurements at the Co L_3 edge on a Pt/Co/GdOx sample [39]. A clear correlation between the evolution of the magnetic anisotropy and the Co oxidation was demonstrated [Fig. 5.12(a,b)]. In Fig. 5.12b, one can see that the shape of the Co XAS spectra are modified with voltage. Whereas a negative voltage induces some fine structures in the XAS spectra indicating a Co oxidation, a positive voltage reduces the Co layer, and therefore Co spectra become similar to that of metallic Co. Thus the Co magnetic anisotropy in this Pt/Co/GdOx structures can be reversibly controlled by voltage via Co oxidation and reduction. This result confirms the earlier experiments performed by F. Bonell et al. [40] on Au/CoFe/MgO samples. Hence, in these examples, the modification of magnetic anisotropy is rather due to oxygen electromigration than charge accumulation.

In V/Fe/MgO devices, S. Miwa et al. [41] have indeed observed a different behaviour. A slight change of coercive field with bias voltage has been measured [see Fig. 5.12c] but without any modification of the Fe $L_{2,3}$ XAS and X-ray magnetic circular dichroism (XMCD) spectra with the applied voltage, suggesting that Fe is not oxidized [Fig. 5.12d]. This is quite surprising since the electric field applied is similar in Pt/Co/GdOx and V/Fe/MgO experiments. This result shows that modifications of the coercive field is not induced by electromigration but is rather due to charge accumulation.

To conclude, this careful investigation of the electronic properties of the FM/oxide interface has allowed to unveil the origin of voltage control of magnetic anisotropy in each NM/FM/oxide structures and thus to discriminate between electromigration and charge effects.

5.2.1.2 Effect of Strain

Strain has been also used to tune magnetic anisotropy and can be very efficient if magnetostrictive materials are used. By combining piezoelectric and magnetostrictive materials, it is possible to control the magnetic anisotropy by voltage. Here we show an example where photoelectron emission microscopy (PEEM) measurements have been performed to highlight the correlation between strain and magnetism [42]. A magnetostrictive $CoFe_2O_4$ layer has been deposited on a piezoelectric $BaTiO_3$ substrate. $BaTiO_3$ can have domains with different unit cell parameters, and therefore can induce different domains with different strains in the $CoFe_2O_4$ layer. X-ray linear dichroism (XLD) being sensitive to local coordination, different XLD spectra will be obtained. In Fig. 5.13d, XLD-PEEM[4] image recorded at the Fe L_3 edge are

---

[4]XAS spectra can be also recorded on the different domains allowing to measure the local electronic properties.

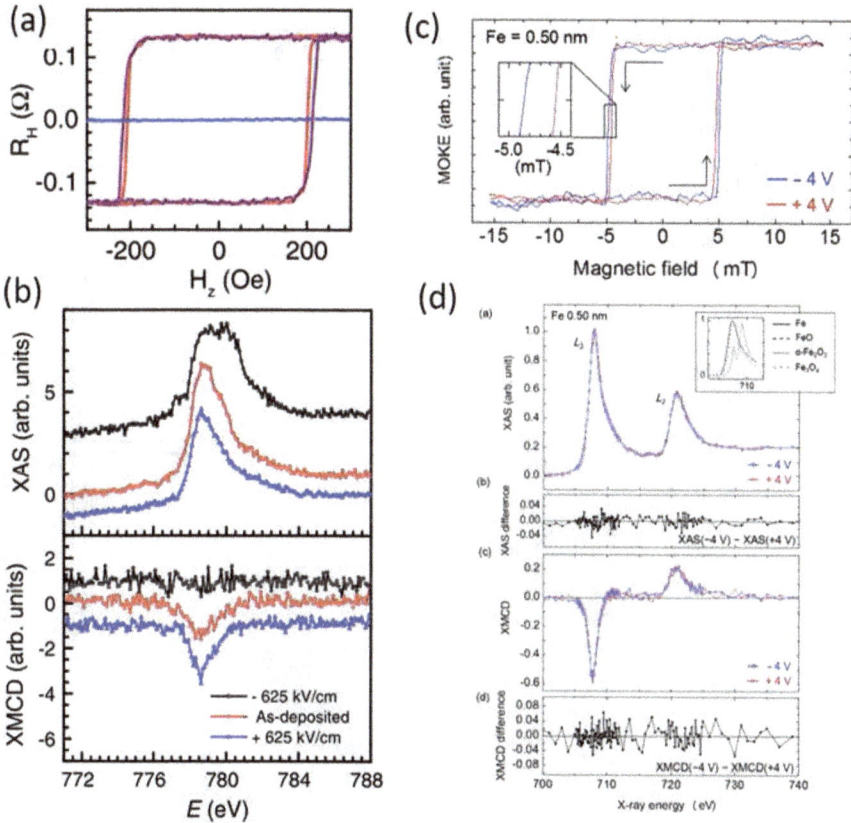

Fig. 5.12 a Hall resistance for a Pt/Co/GdOx sample as deposited (red) and after an applied bias voltage leading to an electric field of -625 kV cm^{-1} (blue) and $+625 \text{ kV cm}^{-1}$ (purple) showing an in-plane to out-of-plane magnetization transition. **b** XAS and XMCD spectra recorded at the Co L_3 edge. A negative voltage induces an oxidation of the Co layer, whereas a positive voltage reduces the Co layer. **c** Magnetic hysteresis loop of a V/Fe/MgO sample obtained for two bias voltages showing a modification of the coercive field. **d** Fe $L_{2,3}$ XAS and XMCD spectra recorded at $+4 \text{ V}$ and -4 V. Fe oxidation/reduction is not observed demonstrating that the coercive field modification is rather attributed to a charge effect. **a, b** Adapted from [39].

presented showing different strain domains for the $CoFe_2O_4$ layer induced by the $BaTiO_3$ substrate. By recording XMCD-PEEM images at the Fe L_3 edge, domains with different magnetic signals are observed. In Fig. 5.13, it appears clearly that a correlation exists between the strain state and the magnetic signal. Black domains in the XLD-PEEM [see Fig. 5.13d] correspond to a magnetization along the [100] direction [see Fig. 5.13a], whereas for white domains, the magnetization is along the [010] direction [see Fig. 5.13c]. Hence, by performing XLD-PEEM and XMCD-

Fig. 5.13 XMCD-PEEM (a–c) and XLD-PEEM (d) images recorded at Fe L_3 edge for a BaTiO$_3$/CoFe$_2$O$_4$ sample. XMCD-PEEM images show magnetic stripes along the [100] and [010] directions. These magnetic stripes are correlated to domains observed in the XLD-PEEM image **d** which indicates different strain states of CoFe$_2$O$_4$ induced by the BaTiO$_3$ substrate.

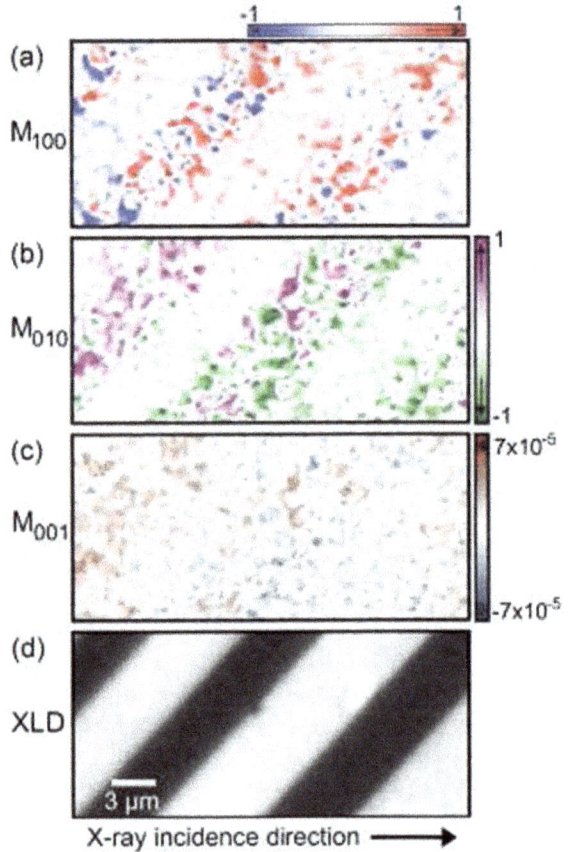

PEEM measurements, it is possible to locally observe the influence of strain on magnetism.

We have shown that XAS has allowed determining the origin of the voltage control of magnetism in NM/FM/oxide structures. In particular, it is possible to disentangle charge and electromigration effects. By performing PEEM measurements with both circular and linear polarized X-rays, the correlation between strain and magnetism can be directly probed. Hence, synchrotron radiation-based measurements are useful to better understand the mechanisms of the voltage control of magnetism.

5.2.2 Spintronics with Pure Spin Current

The generation of pure spin current from heat, charge current, light or vibration is an active and promising research field [43]. A wide variety of materials (ferromagnets, heavy metals, semiconductors, insulators, topological insulators, etc.) and devices are currently investigated. In this paragraph, we aim at discussing the spin-charge conversion using topological insulators as well as the heat-spin current conversion in ferromagnetic insulator/paramagnetic metal devices.

5.2.2.1 Spin-Charge Conversion

Conversion between charge and spin currents has been a very active branch of spintronics in the last couple of years. Such a conversion can be achieved in bulk materials (the so-called spin Hall effect) or at interfaces (Edelstein–Rashba effect) relying on the spin–orbit interaction and/or extrinsic effects. The studied materials are usually NM metals with a large spin–orbit coupling. A second approach is to rely on spin–orbit properties at the interfaces either at Rashba interfaces or through the surface states of topological insulators (TIs). A complete review of this approach can be found in J. Sinova et al. [44] or A. Soumyanarayanan et al. [36].

In the following, we provide an example where angle-resolved photoemission spectroscopy (ARPES) measurements have allowed to understand the spin-charge conversion from the α-Sn TI [see Fig. 5.14a]. In this spin-charge conversion study, the spin current is generated through the magnetization dynamics induced at the magnetization resonance of a Fe layer by an external *rf* field. The generated spin current then diffuses to the α-Sn top surface. Injection of a spin current into a TI induces a spin accumulation on one side of the Fermi contour of the Dirac cone as well as a spin depletion on the other side [see Fig. 5.14b]. As a consequence, this spin injection results in a charge current. Importantly, note that the spin-charge conversion is not observed when Fe is deposited directly on α-Sn, but is observed when a thin Ag layer is inserted at the interface [see Fig. 5.14c]. By performing ARPES measurement to probe the DOS, it has been shown that the deposition of a sub-monolayer of Fe on α-Sn indeed suppresses the Dirac cone, that is a signature of the TI, whereas it is still observable after deposition of a Ag layer [see Fig. 5.14d]. Hence the absence of spin-charge conversion at Fe/α-Sn interfaces is clearly ascribed to the loss of TI surface states after the Fe deposition. This study shows that characterization of the DOS by ARPES measurements is a very useful and unique technique to understand spin-charge conversion at such spinorbitronic interfaces.

5.2.2.2 Heat-Spin Conversion

Similarly to the Seebeck effect, the spin Seebeck effect describes the generation of a spin voltage from a temperature gradient in a FM conductor or insulator. Longitudinal spin Seebeck effect (LSSE) refers to experiments where the spin current generated is parallel to the temperature gradient. Materials used are FM materials (conducting

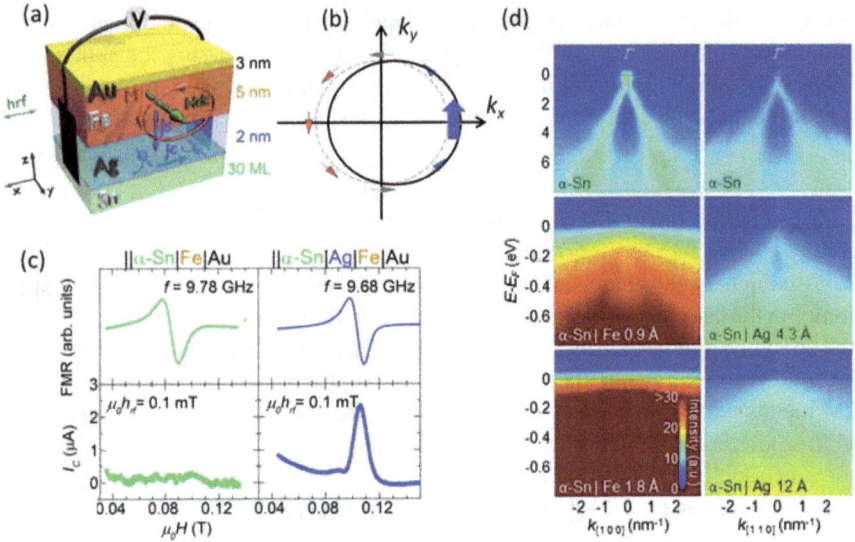

Fig. 5.14 a Scheme of the device studied for the spin to charge conversion by spin pumping into the topological insulator α-Sn. **b** Illustration of the inverse Edelstein effect. A spin current injected in the topological insulator induces an accumulation of charge for one spin direction inducing a shift of the Fermi contour creating a charge current. **c** Ferromagnetic resonance and dc charge current signals for α-Sn/Fe/Au and α-Sn/Ag/Fe/Au samples. Only the second sample shows a dc current signal. **d** ARPES measurement along the [100] or [110] directions on the free surface of α-Sn (top) and after deposition of Fe (left) or Ag (right). The Dirac cone subsists after deposition of Ag but not for Fe.

or insulating) in contact with heavy metals. In order to extract LSSE efficiency, it is necessary to discard the possible artefacts such as an anomalous Nernst effect that arises from the potential induced magnetization in the NM layer [46].

Prototype material systems contain a thin film of yttrium iron garnet $Y_3Fe_5O_{12}$ (YIG) as a FM insulator and a thin layer of Pt as high spin-orbit non-FM material. In order to detect and quantify the magnetic proximity effect (MPE) at the YIG/Pt interface, XMCD measurements appear to be the most accurate method. In Fig. 5.15, XMCD measurements at the Pt $L_{2,3}$ edges performed on YIG/Pt samples are shown. Whereas no MPE have been detected for one study [see Fig. 5.15a] [47, 48], a clear XMCD signal has been observed for another experiment [Fig. 5.15b] [49]. This discrepancy can be attributed to different interface qualities. Actually, x-ray absorption near-edge spectra (XANES) have clearly different shapes. The white line intensity (XANES step height) depends on the number of holes in the $5d$ band and is, therefore, sensitive to the oxidation state of Pt. In the case where no MPE has been detected, the white line intensity is about 1.3 which is similar to the Pt metal value. On the contrary, when MPE is observed, the white line intensity is about 1.45 which is close to what is observed for $PtO_{1.36}$ samples [50]. These observations mean that the observed MPE is probably due to intermixing at the YIG/Pt interface.

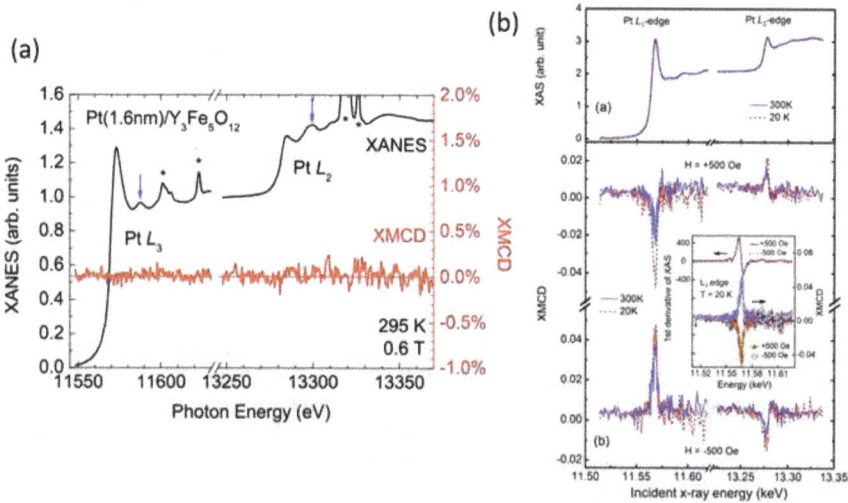

Fig. 5.15 XANES and XMCD recorded at the Pt $L_{2,3}$ edges for YIG/Pt samples. For apparently similar samples a clear XMCD signal is observed for one sample (right), whereas no induced magnetic moment is measured for the other (left). The higher intensity of the step edge observed for the sample exhibiting an XMCD signal (right) meaning an increase of the number of holes in the $5d$ band indicates a possible Pt oxidation. Hence different interface qualities induce different XMCD signals.

This behaviour has been clearly evidenced by further experiments on insulating ferrite/Pt samples. In Fig. 5.16 (left panel), XMCD measurements performed at the Pt M_3 edge on $CoFe_2O_4$/Pt samples are presented [51]. A clear XMCD is observed for samples where Pt has been deposited at high temperature (HT in Fig. 5.16), whereas no XMCD signal is detected when Pt is grown at room temperature (RT in Fig. 5.16). By performing XAS and XMCD measurements at Co and Fe $L_{2,3}$ edges [see Fig. 5.16 (right panel)] it appears clearly that the deposition of Pt at high temperature induces some intermixing at the $CoFe_2O_4$/Pt interface. Indeed, Co XAS spectra are close to those of metallic Co films for the sample with Pt grown at high temperature, whereas for the sample with Pt deposited at room temperature, spectra shapes are similar to those of $CoFe_2O_4$ thin films. Hence the presence of MPE in ferrite/Pt interface comes from interface alloying, whereas for clean interfaces, the induced magnetization by proximity effect is absent. In fact, XAS is not the only measurement to detect MPE; X-ray resonant magnetic reflectivity is also a powerful technique. Thanks to the latter, T. Kuschel et al. have confirmed that no MPE occurs at the clean $NiFe_2O_4$/Pt interface [52]. Finally, it has been shown that no MPE occurs for another ferrite ($MnFe_2O_4$/Pt) and for magnetite, both in the conducting and insulating states [53]. This suggests that the absence of MPE at the magnetic oxide/Pt clean interface is a general rule.

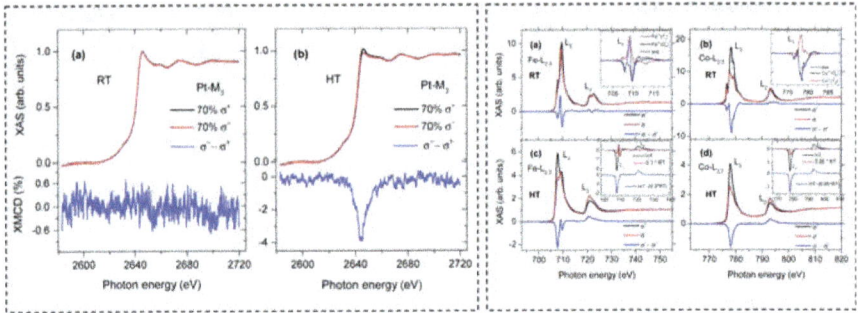

Fig. 5.16 XAS and XMCD spectra recorded at the Pt M_3 edge for $CoFe_2O_4$/Pt where Pt is grown at room temperature (RT) and high temperature (HT). A clear XMCD signal is observed at the Pt M_3 edge when Pt is grown at high temperature. A clear difference can be also seen in the Fe and Co $L_{2,3}$ edges. For Pt grown at HT Co $L_{2,3}$, spectra look like Co metal signifying an intermixing at the $CoFe_2O_4$/Pt interface. When Pt is grown at RT, Co and Fe $L_{2,3}$ XAS spectra are similar to those of $CoFe_2O_4$ thin films.

These experiments hence highlight that synchrotron radiation-based spectroscopies are powerful to probe the electronic and magnetic properties of interfaces and can be very useful to discard artefacts in LSSE experiments.

5.2.3 Current-Induced Magnetization Dynamics

Besides STT-MRAMs, other conceptual memory devices have been proposed and largely studied in the decade. A flagship example is the domain racetrack memory, proposed by S. S. P. Parkin [54], which is a nanoscale shift register memory in which bits are defined by magnetic domains separated by domain walls. Another version of such devices has been recently introduced where the bits are stored thanks to the presence of magnetic skyrmions [55]. In this section, we show that magnetization dynamics imaging using synchrotron radiation-based measurements that can be useful to better understand these STT-MRAM and racetrack memory devices.

5.2.3.1 Spin–Orbit Torque Driven Magnetization Reversal

SOT is an efficient tool to manipulate magnetization in spintronic devices integrating only one FM electrode (see, for example, R. Ramaswamy et al. for a recent review [56]). M. Baumgartner et al. [57] have studied Co nanodot magnetization reversal by SOT. In Fig. 5.17a, the magnetization reversal [measured by XMCD at the Co L_3 edge (black)] induced by a current pulse (red) is shown. Then, by performing time-resolved scanning transmission X-ray microscopy (STXM) measurements, it has

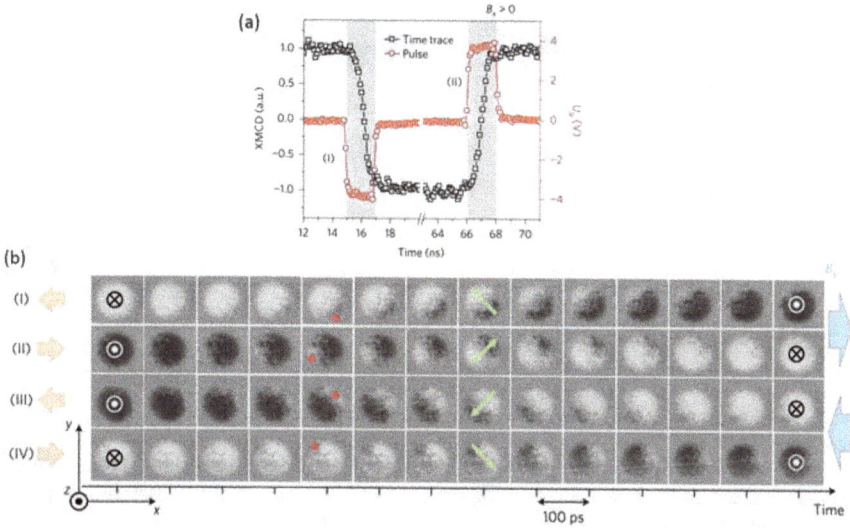

Fig. 5.17 **a** Magnetization reversal probed by XMCD measurements at Co L_3 edge (black) induced by a current pulse (red). **b** STXM images recorded at intervals of 100 ps during a 2 ns injected current pulse. The four rows correspond to different applied current (yellow arrows) and magnetic field (blue arrows) conditions. Red dots indicate domain wall nucleation and green arrows the direction of the domain wall propagation.

been possible to perform a direct observation of the actual path leading to the magnetization reversal during the current pulse injection. Figure 5.17b represents STXM images recorded at the Co L_3 edge with 25 nm spatial resolution at intervals of 100 ps during a 2 ns current pulse. From these images, nucleation (red dot) and propagation (green arrow) of magnetic domain walls can be clearly observed. Depending on the applied magnetic field direction and the current polarity, the main characteristics of these nucleation and propagation might change. These measurements have allowed demonstrating that this diagonal motion of the domain wall originates from a combination of the damping-like and field-like SOTs and the Dzyaloshinskii–Moriya interaction.

5.2.3.2　Current-Induced Domain Walls and Skyrmions Motion

Racetrack memory-based on domain wall or skyrmion motion is an attractive field for future magnetic memories. Synchrotron radiation-based techniques such as PEEM and STXM or even X-ray magnetic resonant scattering (XMRS) have allowed to study the fine structures of magnetic textures such as chiral domain walls [58], vortex [59] and skyrmions [60, 61], see, for example, X. Cheng et al. for a review [62]. A key point is also to optimize the motion velocity and determine the factors that could

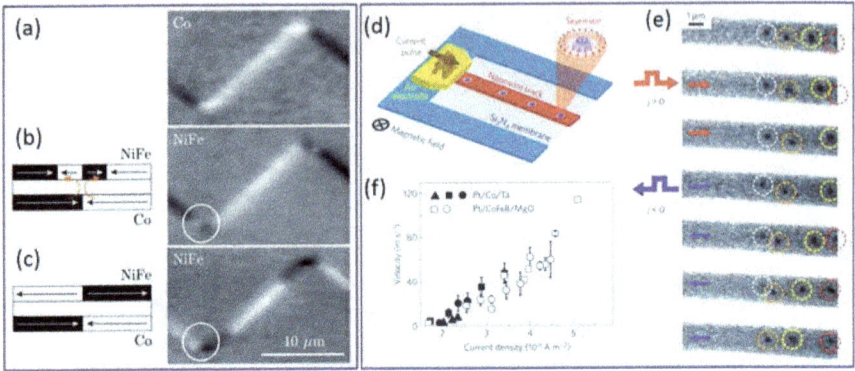

Fig. 5.18 (left panel) XMCD images of Co and NiFe electrodes of a NiFe/Cu/Co spin valves recorded at the Co L_3 edge **a** and Fe L_3 edge **b**, **c**. Different magnetic domains are observed depending on the parallel **b** or antiparallel **c** magnetic configuration (right panel). **d** Sketch of the Pt/Co/Ta stripe deposited on a Si_3N_4 membrane. **e** STXM images recorded after current pulses showing that three of the four skyrmions move (orange, yellow and red circles) after the current pulse. The fourth skyrmion (white circle) is pinned. **f** Skyrmion velocity as a function current density injected for Pt/Co/Ta and Pt/CoFeB/MgO devices.

limit this velocity. X-ray imaging is a powerful tool to investigate magnetic object motion induced by a current. In the following, we show two examples of domain wall and skyrmion motions probed by XMCD-PEEM and STXM.

Large domain wall velocities ($\sim$600 m s^{-1}) have been measured in NiFe/Cu/Co spin valves [63]. Unexpectedly, the domain wall motion is altered when longer electrical pulses or higher current densities are applied [63]. By performing XMCD-PEEM measurements, it has been shown that the dipolar interaction between the NiFe and Co electrodes was a source of domain wall pinning. Thanks to the chemical selectivity of X-ray photoemission, magnetic configuration of the NiFe and Co electrodes can be probed by recording images at the Fe and Co L_3 edges [see Fig. 5.18a, c]. In the parallel magnetic configuration, the stray field of the Co domain wall locally reverses the magnetization in the NiFe layer, leading to the three domain walls [white circle in Fig. 5.18b]. On the other hand, in the antiparallel magnetic configuration, the magnetic flux closes naturally and a single domain wall is formed [Fig. 5.18c]. This magnetic imaging of the Co and NiFe electrodes demonstrates that the stray field prevents domain wall motion across the corners in the NiFe layer.

An important property of the racetrack memories is that the magnetic objects (domain walls or skyrmions) move all together. In Fig. 5.18(d-f), we show that STXM measurements can be used to probe current-driven skyrmion motion in a Pt/Co/Ta stripe as shown by K. Woo *et al.* [29]. Each image is recorded after injection of current pulses. Three of the four skyrmions (red, yellow and orange circles) move forward and backward depending on the current polarity [see Fig. 5.18e]. Note that the fourth skyrmion (white circle) is not showing any motion under current injection (for both polarities) and is probably pinned by a material defect. Skyrmion velocities can be extracted [see Fig. 5.18f] and it is shown that they increase from $\sim 50\,\text{m}\,\text{s}^{-1}$ to $\sim 120\,\text{m}\,\text{s}^{-1}$ by replacing Co by CoFeB. These measurements allow to conclude that lower pinning materials such as amorphous CoFeB layer, i.e. without grain boundaries, are probably more appropriate to get an efficient skyrmion motion.

X-ray microscopy is thus a powerful tool to study in real time the current-driven magnetization dynamics in various magnetic systems. We have shown three examples where magnetic reversal, domain walls and skyrmion motion are investigated by STXM or XMCD-PEEM. This is not limited to FM systems as antiferromagnetic domains can be also probed by performing linear dichroism. Recently, M. J. Grzybowski et al. have measured current-induced antiferromagnetic domain switching in CuMnAs by performing XMLD-PEEM measurements [64].

5.3 Conclusion

We have shown through few examples that synchrotron radiation-based spectroscopies are powerful techniques to perform advanced characterization of various spintronic systems. For example, XAS (including circular and linear dichroism) allows probing the electronic and magnetic properties of surfaces, interfaces and bulk materials. Spin- and angle-resolved photoemission is an ideal tool to measure the actual spin polarization and reveal the density of states of surfaces and interfaces. It can be applied to a wide variety of materials used in spintronic devices such as ferromagnets, antiferromagnets, topological insulators, Rashba interfaces, hybrid ferromagnet/molecules interfaces, 2D materials and multiferroics. XPEEM and STXM are also perfect techniques to study the dynamics of magnetic structures (ferromagnetic, ferrimagnetic and antiferromagnetic) but also to measure locally the electronic structure of surfaces. This is not exhaustive; other techniques such as X-ray magnetic scattering, resonant inelastic X-ray scattering, hard X-ray photoelectron spectroscopy, nano-ARPES can be also used to probe spintronic device properties. Finally, the development of new techniques such as magnetic X-ray nanotomography [65] and novel X-ray sources (new generation of synchrotron and X-ray free-electron lasers) improving coherence, spatial and temporal resolution, will allow obtaining deeper characterization and understanding of magnetic textures and spintronic devices.

Acknowledgements Financial support from the Agence Nationale de la Recherche, France, under grant agreement No. ANR-17-CE24-0025 (TOPSKY), the DARPA TEE program, through grant MIPR No. HR0011831554 and the Horizon2020 Framework Programme of the European Commission, under FET-Proactive Grant agreement No. 824123 (SKYTOP), is acknowledged.

References

1. N.F. Mott, The electrical conductivity of transition metals. Proc. R. Soc. Lond., Ser. A **153**, 699 (1936). https://doi.org/10.1098/rspa.1936.0031
2. A. Fert, I. Campbell, Two-current conduction in nickel. Phys. Rev. Lett. **21**, 1190 (1968). https://doi.org/10.1103/PhysRevLett.21.1190
3. A. Fert, I. Campbell, Transport properties of ferromagnetic transition metals. J. Phys. **32** C1, 46 (1971). https://doi.org/10.1051/jphyscol:1971109
4. A. Fert, I. Campbell, Electrical resistivity of ferromagnetic nickel and iron based alloys. J. Phys. F: Metal Phys. **6**, 849 (1976). https://doi.org/10.1088/0305-4608/6/5/025
5. A. Fert, Nobel lecture: origin, development, and future of spintronics. Rev. Mod. Phys. **80**, 1517 (2008). https://doi.org/10.1103/RevModPhys.80.1517
6. P. Grünberg, R. Schreiber, Y. Pang, M. Brodsky, H. Sowers, Layered magnetic structures: evidence for antiferromagnetic coupling of Fe layers across Cr interlayers. Phys. Rev. Lett. **57**, 2442 (1986). https://doi.org/10.1103/PhysRevLett.57.2442
7. B. Dieny, V.S. Speriosu, S.S.P. Parkin, B.A. Gurney, D.R. Wilhoit, D. Mauri, Giant magnetoresistive in soft ferromagnetic multilayers. Phys. Rev. B **43**, 1297 (1991). https://doi.org/10.1103/PhysRevB.43.1297
8. A. Barthélémy, A. Fert, F. Petroff, Giant magnetoresistance in magnetic multilayers, in *Handbook of Magnetic Materials*, ed. by K.H.J. Buschow, vol. 12 (North-Holland, Amsterdam, 1999), p. 1. https://doi.org/10.1016/S1567-2719(99)12005-5
9. M.N. Baibich, J.M. Broto, A. Fert, F.N. Van Dau, F. Petroff, P. Etienne, G. Creuzet, A. Friederich, J. Chazelas, Giant magnetoresistance of (001)Fe/(001)Cr magnetic superlattices. Phys. Rev. Lett. **61**, 2472 (1988). https://doi.org/10.1103/PhysRevLett.61.2472
10. G. Binasch, P. Grünberg, F. Saurenbach, W. Zinn, Enhanced magnetoresistance in layered magnetic structures with antiferromagnetic interlayer exchange. Phys. Rev. B **39**, 4828 (1989). https://doi.org/10.1103/PhysRevB.39.4828
11. T. Valet, A. Fert, Theory of the perpendicular magnetoresistance in magnetic multilayers. Phys. Rev. B **48**, 7099 (1993). https://doi.org/10.1103/PhysRevB.48.7099
12. M. Jullière, Tunneling between ferromagnetic films. Phys. Lett. A **54**, 225 (1975). https://doi.org/10.1016/0375-9601(75)90174-7
13. J.S. Moodera, L.R. Kinder, T.M. Wong, R. Meservey, Large magnetoresistance at room temperature in ferromagnetic thin film tunnel junctions. Phys. Rev. Lett. **74**, 3273 (1995). https://doi.org/10.1103/PhysRevLett.74.3273
14. T. Miyazaki, N. Tezuka, Giant magnetic tunneling effect in Fe/Al$_2$O$_3$/Fe junction. J. Magn. Magn. Mater. **139**, L231 (1995). https://doi.org/10.1016/0304-8853(95)90001-2
15. M. Bowen, V. Cros, F. Petroff, A. Fert, C. Martínez Boubeta, J.L. Costa-Krämer, J.V. Anguita, A. Cebollada, F. Briones, J.M. de Teresa, L. Morellón, M.R. Ibarra, F. Güell, F. Peiró, A. Cornet, Large magnetoresistance in Fe/MgO/FeCo(001) epitaxial tunnel junctions on GaAs(001). Appl. Phys. Lett. **79**, 1655 (2001). https://doi.org/10.1063/1.1404125
16. J. Faure-Vincent, C. Tiusan, E. Jouguelet, F. Canet, M. Sajieddine, C. Bellouard, E. Popova, M. Hehn, F. Montaigne, A. Schuhl, High tunnel magnetoresistance in epitaxial Fe/MgO/Fe tunnel junctions. Appl. Phys. Lett. **82**, 4507 (2003). https://doi.org/10.1063/1.1586785
17. W.H. Butler, X.-G. Zhang, T.C. Schulthess, J.M. MacLaren, Spin-dependent tunneling conductance of Fe|MgO|Fe sandwiches. Phys. Rev. B **63**, 054416 (2001). https://doi.org/10.1103/PhysRevB.63.054416
18. S. Yuasa, D. Djayaprawira, Giant tunnel magnetoresistance in magnetic tunnel junctions with a crystalline MgO(001) Barrier. J. Phys. D: Appl. Phys. **40**, R337 (2007). https://doi.org/10.1088/0022-3727/40/21/R01
19. S. Ikeda, J. Hayakawa, Y. Ashizawa, Y.M. Lee, K. Miura, H. Hasegawa, M. Tsunoda, F. Matsukura, H. Ohno, Tunnel magnetoresistance of 604% at 300 K by suppression of Ta diffusion in CoFeB/MgO/CoFeB pseudo-spin-valves annealed at high temperature. Appl. Phys. Lett. **93**, 082508 (2008). https://doi.org/10.1063/1.2976435

20. S. Yuasa, T. Nagahama, A. Fukushima, Y. Suzuki, K. Ando, Giant room-temperature magne-toresistance in single-crystal Fe/MgO/Fe magnetic tunnel junctions. Nat. Mater. **3**, 868 (2004). https://doi.org/10.1038/nmat1257

21. S.S.P. Parkin, C. Kaiser, A. Panchula, P.M. Rice, B. Hughes, M. Samant, S.-H. Yang, Giant tunnelling magnetoresistance at room temperature with MgO(100) tunnel barriers. Nat. Mater. **3**, 862 (2004). https://doi.org/10.1038/nmat1256

22. J.C. Slonczewski, Current-driven excitation of magnetic multilayers. J. Magn. Magn. Mater. **159**, L1 (1996). https://doi.org/10.1016/0304-8853(96)00062-5

23. L. Berger, Emission of spin waves by a magnetic multilayer traversed by a current. Phys. Rev. B **54**, 9353 (1996). https://doi.org/10.1103/PhysRevB.54.9353

24. M. Tsoi, A.G.M. Jansen, J. Bass, W.-C. Chiang, M. Seck, V. Tsoi, P. Wyder, Excitation of a magnetic multilayer by an electric current. Phys. Rev. Lett. **80**, 4281 (1998). [Erratum: Phys. Rev. Lett. **81**, 493 (1998).] https://doi.org/10.1103/PhysRevLett.80.4281, [https://doi.org/10.1103/PhysRevLett.81.493]

25. J.A. Katine, F.J. Albert, R.A. Buhrman, E.B. Myers, D.C. Ralph, Current-driven magnetization reversal and spin-wave excitations in Co/Cu/Co pillars. Phys. Rev. Lett. **84**, 3149 (2000). https://doi.org/10.1103/PhysRevLett.84.3149

26. Z. Diao, D. Apalkov, M. Pakala, Y. Ding, A. Panchula, Y. Huai, Spin transfer switching and spin polarization in magnetic tunnel junctions with MgO and AlO$_x$ barriers. Appl. Phys. Lett. **87**, 232502 (2005). https://doi.org/10.1063/1.2139849

27. J. Grollier, D. Lacour, V. Cros, A. Hamzic, A. Vaures, A. Fert, D. Adam, G. Faini, Switching the magnetic configuration of a spin valve by current-induced domain wall motion. J. Appl. Phys. **92**, 4825 (2002). https://doi.org/10.1063/1.1507820

28. A. Dussaux, B. Georges, J. Grollier, V. Cros, A. Khvalkovskiy, A. Fukushima, M. Konoto, H. Kubota, K. Yakushiji, S. Yuasa, K.A. Zvezdin, K. Ando, A. Fert, Large microwave generation from current-driven magnetic vortex oscillators in magnetic tunnel junctions. Nat. Commun. **1**, 8 (2010). https://doi.org/10.1038/ncomms1006

29. S. Woo, K. Litzius, B. Krüger, M.-Y. Im, L. Caretta, K. Richter, M. Mann, A. Krone, R.M. Reeve, M. Weigand, P. Agrawal, I. Lemesh, M.-A. Mawass, P. Fischer, M. Kläui, G.S.D. Beach, Observation of room-temperature magnetic skyrmions and their current-driven dynamics in ultrathin metallic ferromagnets. Nat. Mater. **15**, 501 (2016). https://doi.org/10.1038/nmat4593

30. W. Legrand, D. Maccariello, N. Reyren, K. Garcia, C. Moutafis, C. Moreau-Luchaire, S. Collin, K. Bouzehouane, V. Cros, A. Fert, Room-temperature current-induced generation and motion of sub-100 nm skyrmions. Nano Lett. **17**, 2703 (2017). https://doi.org/10.1021/acs.nanolett.7b00649

31. C. Song, B. Cui, F. Li, X. Zhou, F. Pan, Recent progress in voltage control of magnetism: materials, mechanisms, and performance. Prog. Mater. Sci. **87**, 33 (2017). https://doi.org/10.1016/j.pmatsci.2017.02.002

32. H. Ohno, D. Chiba, F. Matsukura, T. Omiya, E. Abe, T. Dietl, Y. Ohno, K. Ohtani, Electric-field control of ferromagnetism. Nature **408**, 944 (2000). https://doi.org/10.1038/35050040

33. T. Dietl, H. Ohno, Dilute ferromagnetic semiconductors: physics and spintronic structures. Rev. Mod. Phys. **86**, 187 (2014). https://doi.org/10.1103/RevModPhys.86.187

34. M. Weisheit, S. Fähler, A. Marty, Y. Souche, C. Poinsignon, D. Givord, Electric field-induced modification of magnetism in thin-film ferromagnets. Science **315**, 349 (2007). https://doi.org/10.1126/science.1136629

35. J. Lou, M. Liu, D. Reed, Y. Ren, N.X. Sun, Giant electric field tuning of magnetism in novel multiferroic FeGaB/lead zinc niobate-lead titanate (PZN-PT) heterostructures. Adv. Mater. **21**, 4711 (2009). https://doi.org/10.1002/adma.200901131

36. A. Soumyanarayanan, N. Reyren, A. Fert, C. Panagopoulos, Emergent phenomena induced by spin-orbit coupling at surfaces and interfaces. Nature **539**, 509 (2016). https://doi.org/10.1038/nature19820

37. A. Manchon, C. Ducruet, L. Lombard, S. Auffret, B. Rodmacq, B. Dieny, S. Pizzini, J. Vogel, V. Uhlíř, M. Hochstrasser, G. Panaccione, Analysis of oxygen induced anisotropy crossover in Pt/Co/*M*Ox trilayers. J. Appl. Phys. **104**, 043914 (2008). https://doi.org/10.1063/1.2969711

38. D. Chiba, S. Fukami, K. Shimamura, N. Ishiwata, K. Kobayashi, T. Ono, Electrical control of the ferromagnetic phase transition in cobalt at room temperature. Nat. Mater. **10**, 853 (2011). https://doi.org/10.1038/nmat3130
39. C. Bi, Y. Liu, T. Newhouse-Illige, M. Xu, M. Rosales, J.W. Freeland, O. Mryasov, S. Zhang, G.E. te Velthuis, W.G. Wang, Reversible control of Co magnetism by voltage-induced oxidation. Phys. Rev. Lett. **113**, 267202 (2014). https://doi.org/10.1103/PhysRevLett.113.267202
40. F. Bonell, Y.T. Takahashi, D.D. Lam, S. Yoshida, Y. Shiota, S. Miwa, T. Nakamura, Y. Suzuki, Reversible change in the oxidation state and magnetic circular dichroism of Fe driven by an electric field at the FeCo/MgO interface. Appl. Phys. Lett. **102**, 152401 (2013). https://doi.org/10.1063/1.4802030
41. S. Miwa, K. Matsuda, K. Tanaka, Y. Kotani, M. Goto, T. Nakamura, Y. Suzuki, Voltage-controlled magnetic anisotropy in Fe|MgO tunnel junctions studied by X-ray absorption spectroscopy. Appl. Phys. Lett. **107**, 162402 (2015). https://doi.org/10.1063/1.4934568
42. R.V. Chopdekar, V.K. Malik, A. Fraile Rodríguez, L. Le Guyader, Y. Takamura, A. Scholl, D. Stender, C.W. Schneider, C. Bernhard, F. Nolting, L.J. Heyderman, Spatially resolved strain-imprinted magnetic states in an artificial multiferroic. Phys. Rev. B **86**, 014408 (2012). https://doi.org/10.1103/PhysRevB.86.014408
43. Y. Otani, M. Shiraishi, A. Oiwa, E. Saitoh, S. Murakami, Spin conversion on the nanoscale. Nat. Phys. **13**, 829 (2017). https://doi.org/10.1038/nphys4192
44. J. Sinova, S.O. Valenzuela, J. Wunderlich, C.H. Back, T. Jungwirth, Spin Hall effects. Rev. Mod. Phys. **87**, 1213 (2015). https://doi.org/10.1103/RevModPhys.87.1213
45. J.C. Rojas-Sánchez, S. Oyarzún, Y. Fu, A. Marty, C. Vergnaud, S. Gambarelli, L. Vila, M. Jamet, Y. Ohtsubo, A. Taleb-Ibrahimi, P. Le Fèvre, F. Bertran, N. Reyren, J.M. George, A. Fert, Spin to charge conversion at room temperature by spin pumping into a new type of topological insulator: α-Sn films. Phys. Rev. Lett. **116**, 096602 (2016). https://doi.org/10.1103/PhysRevLett.116.096602
46. S.Y. Huang, X. Fan, D. Qu, Y.P. Chen, W.G. Wang, J. Wu, T.Y. Chen, J.Q. Xiao, C.L. Chien, Transport magnetic proximity effects in platinum. Phys. Rev. Lett. **109**, 107204 (2012). https://doi.org/10.1103/PhysRevLett.109.107204
47. S. Geprägs, S. Meyer, S. Altmannshofer, M. Opel, F. Wilhelm, A. Rogalev, R. Gross, S.T. Goennenwein, Investigation of induced Pt magnetic polarization in Pt/$Y_3Fe_5O_{12}$ bilayers. Appl. Phys. Lett. **101**, 262407 (2012). https://doi.org/10.1063/1.4773509
48. S. Geprägs, S.T.B. Goennenwein, M. Schneider, F. Wilhelm, K. Ollefs, A. Rogalev, M. Opel, R. Gross, Comment on "Pt magnetic polarization on $Y_3Fe_5O_{12}$ and magnetotransport characteristics" (2013), arXiv:1307.4869v1 [cond-mat.mtrl-sci]
49. Y.M. Lu, Y. Choi, C.M. Ortega, X.M. Cheng, J.W. Cai, S.Y. Huang, L. Sun, C.L. Chien, Pt magnetic polarization on $Y_3Fe_5O_{12}$ and magnetotransport characteristics. Phys. Rev. Lett. **110**, 147207 (2013). https://doi.org/10.1103/PhysRevLett.110.147207
50. A.V. Kolobov, F. Wilhelm, A. Rogalev, T. Shima, J. Tominaga, Thermal decomposition of sputtered thin PtO_x layers used in super-resolution optical disks. Appl. Phys. Lett. **86**, 121909 (2005). https://doi.org/10.1063/1.1886255
51. H.B. Vasili, M. Gamino, J. Gàzquez, F. Sánchez, M. Valvidares, P. Gargiani, E. Pellegrin, J. Fontcuberta, Magnetoresistance in hybrid Pt/$CoFe_2O_4$ bilayers controlled by competing spin accumulation and interfacial chemical reconstruction. ACS Appl. Mater. Interfaces **10**, 12031 (2018). https://doi.org/10.1021/acsami.8b00384
52. Static magnetic proximity effect in Pt/$NiFe_2O_4$ and Pt/Fe bilayers investigated by X-ray resonant magnetic reflectivity. Phys. Rev. Lett. **115**, 097401 (2015). https://doi.org/10.1103/PhysRevLett.115.097401
53. M. Collet, R. Mattana, J.-B. Moussy, K. Ollefs, S. Collin, C. Deranlot, A. Anane, V. Cros, F. Petroff, F. Wilhelm, A. Rogalev, Investigating magnetic proximity effects at ferrite/Pt interfaces. Appl. Phys. Lett. **111**, 202401 (2017). https://doi.org/10.1063/1.4987145
54. S.S.P. Parkin, M. Hayashi, L. Thomas, Magnetic domain-wall racetrack memory. Science **320**, 190 (2008). https://doi.org/10.1126/science.1145799

55. A. Fert, V. Cros, J. Sampaio, Skyrmions on the track. Nat. Nanotechnol. **8**, 152 (2013). https://doi.org/10.1038/nnano.2013.29

56. R. Ramaswamy, J.M. Lee, K. Cai, H. Yang, Recent advances in spin-orbit torques: moving towards device applications. Appl. Phys. Rev. **5**, 031107 (2018). https://doi.org/10.1063/1.5041793

57. M. Baumgartner, K. Garello, J. Mendil, C.O. Avci, E. Grimaldi, C. Murer, J. Feng, M. Gabureac, C. Stamm, Y. Acremann, S. Finizio, S. Wintz, J. Raabe, P. Gambardella, Spatially and time-resolved magnetization dynamics driven by spin-orbit torques. Nat. Nanotechnol. **12**, 980 (2017). https://doi.org/10.1038/nnano.2017.151

58. J.-Y. Chauleau, W. Legrand, N. Reyren, D. Maccariello, S. Collin, H. Popescu, K. Bouzehouane, V. Cros, N. Jaouen, A. Fert, Chirality in magnetic multilayers probed by the symmetry and the amplitude of dichroism in X-ray resonant magnetic scattering. Phys. Rev. Lett. **120**, 037202 (2018). https://doi.org/10.1103/PhysRevLett.120.037202

59. T. Taniuchi, M. Oshima, H. Akinaga, K. Ono, Vortex-chirality control in mesoscopic disk magnets observed by photoelectron emission microscopy. J. Appl. Phys. **97**, 10J904 (2005). https://doi.org/10.1063/1.1862032

60. C. Moreau-Luchaire, C. Moutafis, N. Reyren, J. Sampaio, C.A.F. Vaz, N. Van Horne, K. Bouzehouane, K. Garcia, C. Deranlot, P. Warnicke, P. Wohlhüter, J.-M. George, M. Weigand, J. Raabe, V. Cros, A. Fert, Additive interfacial chiral interaction in multilayers for stabilization of small individual skyrmions at room temperature. Nat. Nanotechnol. **11**, 444 (2016). https://doi.org/10.1038/nnano.2015.313

61. O. Boulle, J. Vogel, H. Yang, S. Pizzini, D. de Souza Chaves, A. Locatelli, T.O. Mentes, A. Sala, L.D. Buda-Prejbeanu, O. Klein, M. Belmeguenai, Y. Roussigné, A. Stashkevich, S.M. Chérif, L. Aballe, M. Foerster, M. Chshiev, S. Auffret, I.M. Miron, G. Gaudin, Room-temperature chiral magnetic skyrmions in ultrathin magnetic nanostructures. Nat. Nanotechnol. **11**, 449 (2016). https://doi.org/10.1038/nnano.2015.315

62. X. Cheng, D. Keavney, Studies of nanomagnetism using synchrotron-based X-ray photoemission electron microscopy (X-PEEM). Rep. Prog. Phys. **75**, 026501 (2012). https://doi.org/10.1088/0034-4885/75/2/026501

63. V. Uhlíř, S. Pizzini, N. Rougemaille, J. Novotný, V. Cros, E. Jiménez, G. Faini, L. Heyne, F. Sirotti, C. Tieg, A. Bendounan, F. Maccherozzi, R. Belkhou, J. Grollier, A. Anane, J. Vogel, Current-induced motion and pinning of domain walls in spin-valve nanowires studied by XMCD-PEEM. Phys. Rev. B **81**, 224418 (2010). https://doi.org/10.1103/PhysRevB.81.224418

64. M.J. Grzybowski, P. Wadley, K.W. Edmonds, R. Beardsley, V. Hills, R.P. Campion, B.L. Gallagher, J.S. Chauhan, V. Novak, T. Jungwirth, F. Maccherozzi, S.S. Dhesi, Imaging current-induced switching of antiferromagnetic domains in CuMnAs. Phys. Rev. Lett. **118**, 057701 (2017). https://doi.org/10.1103/PhysRevLett.118.057701

65. C. Donnelly, M. Guizar-Sicairos, V. Scagnoli, S. Gliga, M. Holler, J. Raabe, L.J. Heyderman, Three-dimensional magnetization structures revealed with X-ray vector nanotomography. Nature **547**, 328 (2017). https://doi.org/10.1038/nature23006

6

d-Vector Representation for Describing p-Wave Superconductivity

Jean-Pascal Brison

Abstract Since the mid-80s, new classes of superconductors have been discovered in which the origin of superconductivity cannot be attributed to the electron–ion interactions at the heart of conventional superconductivity. Most of these unconventional superconductors are strongly correlated electron systems, and identifying (or even more difficult, predicting) the precise superconducting state has been, and sometimes remains, an actual challenge. However, in most cases, it has been demonstrated that in these materials the spin state of the Cooper pairs is a singlet state, often associated with a 'd-wave' or '$s + /-$' orbital state. For a few systems, a spin-triplet state is strongly suspected, like in superfluid ^{3}He; this leads to a much more complex superconducting order parameter. This was long supposed to be the case for the d-electron system Sr_2RuO_4, and is very likely realized in some uranium-based (f-electron) 'heavy fermions' like UPt_3 (with multiple superconducting phases) or UGe_2 (with coexisting ferromagnetic order). Beyond the interest for these materials, p-wave superconductivity is presently quite fashionable for its topological properties and the prediction that it could host Majorana-like low energy excitations, seen as a route towards robust (topologically protected) qubits. The aim of these notes is to make students and experimentalists more familiar with the **d**-vector representation used to describe p-wave (spin triplet) superconductivity. The interest of this formalism will be illustrated on some systems where p-wave superconductivity is the prime suspect.

6.1 Introduction

The purpose of these notes is only to cover some aspects of spin-triplet superconductors, not so commonly covered in the excellent textbooks available on superconductivity, in general, and unconventional superconductors, in particular. Among

J.-P. Brison (✉)
CEA, IRIG, Pheliqs, Université Grenoble Alpes, 17 avenue des Martyrs, 38054 Grenoble Cedex 09, France
e-mail: jean-pascal.brison@cea.fr

those, let us choose to quote only two: the seminal Reviews of Modern Physics paper "A theoretical description of the new phases of liquid ^{3}He" by A. J. Leggett [1], which gives both very advanced and detailed insights on the theory of the p-wave order parameter of superfluid ^{3}He, and pedagogical and enlightening treatment of the microscopic Bardeen–Cooper–Schrieffer (BCS) theory of anisotropic superconductors and the other, which covers the very important symmetry aspects of unconventional superconductors in crystalline materials, is the book 'Introduction to unconventional superconductivity' by V. P. Mineev and K. V. Samokhin [2].

In the following, we concentrate on some basic aspects of the description of spin-triplet superconductors, which are often bewildering, at least to experimentalists.

6.2 Odd-Parity Pairing: BCS Wave Function and Order Parameter

Most known superconductors are 'spin-singlet' superconductors, meaning that the relative wave function of the Cooper pairs $|\Psi(\mathbf{r_1} - \mathbf{r_2})\rangle$, in the real or in the reciprocal space, can be written as a product of an orbital wave function and a spin (singlet) wave function

$$|\Psi(\mathbf{r_1} - \mathbf{r_2})\rangle = \phi(\mathbf{r_1} - \mathbf{r_2})|\uparrow\downarrow - \downarrow\uparrow\rangle ,$$
$$|\Psi(\mathbf{k})\rangle = \varphi(\mathbf{k})|\uparrow\downarrow - \downarrow\uparrow\rangle .$$

(6.1)

Antisymmetrization of the total pair wave function imposes, for such a singlet state, that the orbital wave function verifies $\phi(\mathbf{r_1} - \mathbf{r_2}) = \phi(\mathbf{r_2} - \mathbf{r_1})$ or $\varphi(\mathbf{k}) = \varphi(-\mathbf{k})$ (even-parity state). However, it is also possible to build Cooper pairs in a triplet spin state (see Fig. 6.1). If all electronic interactions including the pairing interactions conserve spin, one could pair separately up- and down-spins, and the total superconducting wave function with a triplet spin state would be the (antisymmetrized) product of both. However, if any non-spin conserving term exists, like the spin–orbit interaction, this is no longer possible. One can just say that Cooper pairs will be formed with a wave function of the form

$$|\Psi\rangle = \phi_{11}(\mathbf{r_1} - \mathbf{r_2})|\uparrow\uparrow\rangle + \phi_{22}(\mathbf{r_1} - \mathbf{r_2})|\downarrow\downarrow\rangle + \phi_{12}(\mathbf{r_1} - \mathbf{r_2})|\uparrow\downarrow + \downarrow\uparrow\rangle , \quad (6.2)$$

or in the reciprocal space

$$|\Psi\rangle = \varphi_{11}(\mathbf{k})|\uparrow\uparrow\rangle + \varphi_{22}(\mathbf{k})|\downarrow\downarrow\rangle + \varphi_{12}(\mathbf{k})|\uparrow\downarrow + \downarrow\uparrow\rangle . \quad (6.3)$$

Antisymmetrization of the total pair wave function imposes this time that the orbital wave function $\phi(\mathbf{r_1} - \mathbf{r_2}) = -\phi(\mathbf{r_2} - \mathbf{r_1})$ or $\varphi(\mathbf{k}) = -\varphi(-\mathbf{k})$ (odd-parity state). Note that microscopically, one would write the ground state superconducting wave function for the whole Fermi sea as

Fig. 6.1 Singlet **a** versus triplet **b** Cooper pairs: they are built with quasiparticles of opposite wave vectors in both cases, but differ by their spin state

$$|\Psi\rangle = \prod_{all\ k} (u_{k\uparrow\uparrow} + v_{k\uparrow\uparrow}c^{+}_{k\uparrow}c^{+}_{-k\uparrow})(u_{k\downarrow\downarrow} + v_{k\downarrow\downarrow}c^{+}_{k\downarrow}c^{+}_{-k\downarrow})(u_{k\uparrow\downarrow} + v_{k\uparrow\downarrow}c^{+}_{k\uparrow}c^{+}_{-k\downarrow})|0\rangle$$

$$= \prod_{all\ k,\alpha,\beta} (u_{k,\alpha\beta} + v_{k,\alpha\beta}c^{+}_{k\alpha}c^{+}_{-k\beta})|0\rangle$$

$$(6.4)$$

with $\quad u_{k,\alpha\beta} = u_{-k,\alpha\beta}$; $v_{k,\alpha\beta} = -v_{-k,\alpha\beta}$,

and $\quad \varphi_{\alpha\beta}(\mathbf{k}) = \langle c_{-k\beta}c_{k\alpha}\rangle = u^{*}_{k\alpha\beta}v_{k\alpha\beta} = -\varphi_{\alpha\beta}(-k)$ the order parameter [1] .

The last condition on the parity of u_k and v_k for the same spin indices ensures that the orbital part is odd (for the exchange of $\mathbf{k}$ and $-\mathbf{k}$), selecting only triplet spin components. Coming back to the order parameter, in the reciprocal space, it should be given by three complex odd functions of $\mathbf{k}$: φ_{11}, φ_{22} and $\varphi_{12} = \varphi_{21}$. The most natural would be to view the order parameter as a 2×2 symmetrical matrix $\varphi_{\alpha\beta}$, where α and β are spin indices ($1 = \uparrow$, $2 = \downarrow$). This is possible, and is used in many calculations. However, it is not very convenient if one needs to change the quantization axis or if (as it commonly happens) the quantization axis changes over the Fermi surface. There are only three independent complex functions of $\mathbf{k}$, so it would be nice to represent the order parameter by a vector.

6.3 Vectors and Cayley–Klein Representation

6.3.1 Position of the Problem

However, this would be meaningful only if this vector transforms properly under rotation of the spin quantization axis. And one would also expect its magnitude to be proportional to the density of condensed Cooper pairs, and its direction to have a meaning relative to the spin orientation. This last point is clearly not so direct, as the vector will necessarily be complex. In order to understand more clearly what is necessary, let us first explore what doesn't work. We could build simply such a vector representation through:

$$\mathbf{V} = \varphi_{11}\mathbf{e_x} + \varphi_{22}\mathbf{e_y} + \varphi_{12}\mathbf{e_z} .$$

$$(6.5)$$

But it would not do the job: the module would be fine, but the direction of $\mathbf{V}$ and so its transformation under rotations of the axis would be meaningless, for example, for the same quantization axis a pure $|\uparrow\uparrow\rangle$ or $|\downarrow\downarrow\rangle$ state would lead to perpendicular vectors. Or equivalently, taking an opposite direction of the quantization axis would yield perpendicular vector representations. This is clearly not what is expected from a vector behaviour. The problem stems from the fact that one needs to make a link between the spin state [SU(2)] and three-dimensional vectors. The good news is that this problem has been solved long ago in classical mechanics, with the Cayley–Klein representation, which aimed at simplifying the calculation of rotation effects; in real space, a matrix rotation is a 3×3 matrix; however, it is fully characterized by only three angles (the Euler angles for example); so, in principle, a 2×2 matrix, with four parameters, should be more than enough. The Cayley–Klein representation associates a three-dimensional vector $(\mathbf{a})$ to a 2×2 matrix through … Pauli matrices

$$\mathbf{a} \rightarrow \mathbf{a}\cdot\boldsymbol{\sigma} ,$$
$$\boldsymbol{\sigma} = \sigma_1 \mathbf{e_x} + \sigma_2 \mathbf{e_y} + \sigma_3 \mathbf{e_z} ,$$
$$\sigma_1 = \sigma_x = \begin{pmatrix} 0 & 1 \\ 1 & 0 \end{pmatrix} \quad \sigma_2 = \sigma_y = \begin{pmatrix} 0 & -i \\ i & 0 \end{pmatrix} \quad \sigma_3 = \sigma_z = \begin{pmatrix} 1 & 0 \\ 0 & -1 \end{pmatrix} , \quad (6.6)$$
$$\sigma_i = \begin{pmatrix} \delta_{3i} & \delta_{1i} - i\delta_{2i} \\ \delta_{1i} + i\delta_{2i} & -\delta_{3i} \end{pmatrix} \quad \mathbf{a}\cdot\boldsymbol{\sigma} = a_i \sigma_i = \begin{pmatrix} a_3 & a_1 - ia_2 \\ a_1 + ia_2 & -a_3 \end{pmatrix} ,$$

where δ_{ij} is the Kronecker symbol.

6.3.2 Useful Formula for Pauli Matrices

As a reminder, for these (Hermitian) Pauli matrices

$$\sigma_i^2 = \mathbb{1} \quad ; \quad [\sigma_i, \sigma_j] = 2i\, \epsilon^{ijk} \sigma_k \quad ; \quad \{\sigma_i, \sigma_j\} = 2\,\delta_{ij}\mathbb{1} ,$$
$$\sigma_i \sigma_j = i\epsilon^{ijk}\sigma_k + \delta_{ij}\mathbb{1} , \quad (6.7)$$
$$\mathrm{tr}(\sigma_i) = 0; \quad \det(\sigma_i) = -1; \quad \text{eigenvalues} = \pm 1 ,$$

where ϵ^{ijk} is the Levi-Civita symbol.

From that, a little algebra leads to very useful formulae ($\mathbf{a}$ and $\mathbf{b}$ are real or complex 3D vectors)

$$(\mathbf{a}\cdot\boldsymbol{\sigma})\sigma_k = (a_i \sigma_i)\sigma_k = a_i \epsilon^{ikj} i\sigma_j + a_i \delta_{ik}\mathbb{1} = -i\epsilon^{kij} a_i \sigma_j + a_k \mathbb{1} .$$

So

$$
\begin{aligned}
&(\mathbf{a}\cdot\boldsymbol{\sigma})\boldsymbol{\sigma} = a\mathbb{1} - i a \wedge \boldsymbol{\sigma} \quad ; \quad \boldsymbol{\sigma}(\mathbf{a}\cdot\boldsymbol{\sigma}) = a\mathbb{1} + i a \wedge \boldsymbol{\sigma} \ ; \\
&tr((\mathbf{a}\cdot\boldsymbol{\sigma})\boldsymbol{\sigma}) = 2\,\mathbf{a} \ ; \\
&(\mathbf{a}\cdot\boldsymbol{\sigma})(\mathbf{b}\cdot\boldsymbol{\sigma}) = (\mathbf{a}\cdot\mathbf{b})\mathbb{1} + i(\mathbf{a}\wedge\mathbf{b})\cdot\boldsymbol{\sigma} \ .
\end{aligned}
\tag{6.8}
$$

Finally, if $\mathbf{a}$ is real, or if at least one can write $\mathbf{a} = a \cdot \hat{\mathbf{a}}$, with a, a complex number, and $\hat{\mathbf{a}}$, a real unit vector, then additional useful relations exist

- the eigenvalues of $\mathbf{a} \cdot \boldsymbol{\sigma}$ are $\pm a$;
- the projectors on each eigenspace can be written as $\frac{1}{2}\left(\mathbb{1} \pm \hat{\mathbf{a}}\cdot\boldsymbol{\sigma}\right)$;
- for any analytic function
  $f(\mathbf{a}\cdot\boldsymbol{\sigma}) = \frac{f(a)}{2}\left(\mathbb{1}+\hat{\mathbf{a}}\cdot\boldsymbol{\sigma}\right) + \frac{f(-a)}{2}\left(\mathbb{1}-\hat{\mathbf{a}}\cdot\boldsymbol{\sigma}\right)$ and
- in particular, if $\boldsymbol{\Omega}$ is a real vector, also written as $\boldsymbol{\Omega} = \Omega\,\hat{\boldsymbol{\Omega}}$, Ω, a real number, and $\hat{\boldsymbol{\Omega}}$, a real unit vector

$$
\exp(i\boldsymbol{\Omega}\cdot\boldsymbol{\sigma}) = \frac{\exp(i\Omega)+\exp(-i\Omega)}{2}\mathbb{1} + i\frac{\exp(i\Omega)-\exp(-i\Omega)}{2i}\hat{\boldsymbol{\Omega}}\cdot\boldsymbol{\sigma} \ ,
$$

$$
\exp(i\boldsymbol{\Omega}\cdot\boldsymbol{\sigma}) = \cos\Omega\,\mathbb{1} + i\sin\Omega\,\hat{\boldsymbol{\Omega}}\cdot\boldsymbol{\sigma} \ .
\tag{6.9}
$$

6.3.3   Rotation of a 3D Vector: Cayley–Klein Relation

From these relations, it is straightforward to see (proof at the end of the chapter) that if $\mathcal{R}$ is a 3D rotation characterized by an angle Ω around the axis $\hat{\boldsymbol{\Omega}}$, for any vector $\mathbf{a}$

$$
\mathcal{R}(\mathbf{a})\cdot\boldsymbol{\sigma} = \exp(-i/2\boldsymbol{\Omega}\cdot\boldsymbol{\sigma})(\mathbf{a}\cdot\boldsymbol{\sigma})\exp(i/2\boldsymbol{\Omega}\cdot\boldsymbol{\sigma}) \ ,
$$

$$
\boxed{\mathcal{R}(\mathbf{a})\cdot\boldsymbol{\sigma} = \mathcal{R}_{\boldsymbol{\Omega}}(\mathbf{a}\cdot\boldsymbol{\sigma})\mathcal{R}_{-\boldsymbol{\Omega}}} \ ,
\tag{6.10}
$$

where $\mathcal{R}_{\boldsymbol{\Omega}} = \exp\left(\frac{-i}{2}\boldsymbol{\Omega}\cdot\boldsymbol{\sigma}\right)$ is the rotation matrix around $\boldsymbol{\Omega}$ for a spin 1/2. So one can work with 2D (complex) matrices to calculate the effect of a 3D rotation $\mathcal{R}$ on a real vector $\mathbf{a}$.

In fact, this is more general in the sense that it is also true when applied on complex vectors (rotating around a real vector $\boldsymbol{\Omega}$). Indeed, the effect of a 3D rotation of angle Ω around the axis $\hat{\boldsymbol{\Omega}}$ on a real vector $\mathbf{a}$ can be easily expressed through the relations (see Fig. 6.2)

$$
\mathbf{a} = (\mathbf{a}\cdot\hat{\boldsymbol{\Omega}})\hat{\boldsymbol{\Omega}} + \mathbf{a} - (\mathbf{a}\cdot\hat{\boldsymbol{\Omega}})\hat{\boldsymbol{\Omega}} \ ,
$$

$$
\boxed{\mathcal{R}(\mathbf{a}) = (\mathbf{a}\cdot\hat{\boldsymbol{\Omega}})\hat{\boldsymbol{\Omega}} + \cos\Omega\left(\mathbf{a} - (\mathbf{a}\cdot\hat{\boldsymbol{\Omega}})\hat{\boldsymbol{\Omega}}\right) + \sin\Omega(\hat{\boldsymbol{\Omega}}\wedge\mathbf{a})} \ .
\tag{6.11}
$$

Fig. 6.2 Decomposition of a vector for the calculation of its rotation by an angle Ω around $\hat{\boldsymbol{\Omega}}$

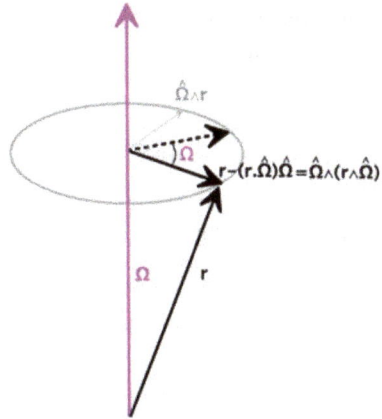

And (6.11) can be used to define what is the rotation of a complex vector around a real vector $\hat{\boldsymbol{\Omega}}$. With such a definition, the Cayley–Klein relation (6.10) also works when $\mathbf{a}$ is a complex 3D vector (see 'proof' in Sect. 6.11).

Exercise 6.1 Show that with the definition of the rotation (6.11) of a complex vector (around a 'real vector $\boldsymbol{\Omega}$'), the scalar product and the cross product are conserved under rotation:

$$\mathcal{R}(\mathbf{d}) \cdot \mathcal{R}(\mathbf{u}) = \mathbf{d}\cdot\mathbf{u} ,$$
$$\mathcal{R}(\mathbf{u}) \wedge \mathcal{R}(\mathbf{d}) = \mathcal{R}(\mathbf{u} \wedge \mathbf{d}) . \tag{6.12}$$

Solution in Sect. 6.11.

6.4 d-Vector Representation

Coming back to the problem of finding a vector representation of the order parameter, if we could cast the 2×2 matrix order parameter (φ) in the form $(\mathbf{a}\cdot\boldsymbol{\sigma})$, there are good chances that the vector $(\mathbf{a})$ would do the job. Working in the reciprocal space $(\mathbf{k} = k_F\hat{\mathbf{n}})$, where k_F is the radius of the Fermi surface, we start from

$$|\Psi(\hat{\mathbf{n}})\rangle = \sum_{\alpha,\beta} \varphi_{\alpha\beta}(\hat{\mathbf{n}})|\alpha\beta\rangle$$
$$= \Delta^{\uparrow}(\hat{\mathbf{n}})|\uparrow\uparrow\rangle + \Delta^{\downarrow}(\hat{\mathbf{n}})|\downarrow\downarrow\rangle + \Delta^{0}(\hat{\mathbf{n}})(|\uparrow\downarrow\rangle + |\downarrow\uparrow\rangle) , \tag{6.13}$$

$$(\varphi) = \begin{pmatrix} \varphi_{\alpha\alpha} & \varphi_{\beta\alpha} \\ \varphi_{\alpha\beta} & \varphi_{\beta\beta} \end{pmatrix} \quad \text{with} \quad \varphi_{\alpha\beta} = \varphi_{\beta\alpha} \quad \text{or} \quad (\varphi) = \begin{pmatrix} \Delta^{\uparrow} & \Delta^{0} \\ \Delta^{0} & \Delta^{\downarrow} \end{pmatrix} . \tag{6.14}$$

Comparison with expression (6.6) for $\mathbf{a \cdot \sigma}$ doesn't fit, notably as regards the non-diagonal symmetric terms. This is due to the σ_2 component of $\mathbf{a \cdot \sigma}$. It can be eliminated if one calculates

$$i(\mathbf{a \cdot \sigma}) \cdot \sigma_2 = i \begin{pmatrix} a_3 & a_1 - ia_2 \\ a_1 + ia_2 & -a_3 \end{pmatrix} \begin{pmatrix} 0 & -i \\ i & 0 \end{pmatrix} . \qquad (6.15)$$

This allows for a straightforward identification of a vector representation of the order parameter, noted as $\mathbf{d}$, of components

$$\varphi_{\alpha\beta} = (i(\mathbf{d} \cdot \boldsymbol{\sigma}) \cdot \sigma_2)_{\alpha\beta} \quad \Longleftrightarrow \quad (\varphi) = i(\mathbf{d} \cdot \boldsymbol{\sigma}) \cdot \sigma_2 . \qquad (6.16)$$

Note that, traditionally, $\mathbf{d}$ is normalized to 1 (in a sense to be precised later), like a wave function, whereas the order parameter amplitude reflects the 'superfluid density' and is proportional to the gap in the simplest cases. The equations above do not reflect this subtlety that will be precised later on (see Sect. 6.6.1). Therefore, to be 'in line' with the convention of most papers on the subject, we will introduce a ($\mathbf{k}$-independent) proportionality factor ψ

$$\left. \begin{aligned} \varphi_{11} &= \Delta^\uparrow = \psi(-d_x + id_y) \\ \varphi_{22} &= \Delta^\downarrow = \psi(d_x + id_y) \\ \varphi_{12} &= \varphi_{21} = \Delta^0 = \psi(d_z) \end{aligned} \right\} \Leftrightarrow \left\{ \begin{aligned} \psi d_x &= \tfrac{1}{2}(-\Delta^\uparrow + \Delta^\downarrow) = \tfrac{1}{2}(-\varphi_{11} + \varphi_{22}) \\ \psi d_y &= -\tfrac{i}{2}(\Delta^\uparrow + \Delta^\downarrow) = -\tfrac{i}{2}(\varphi_{11} + \varphi_{22}) \\ \psi d_z &= \Delta^0 = \tfrac{1}{2}(\varphi_{12} + \varphi_{21}) \end{aligned} \right. \qquad (6.17)$$

And convenient expressions for calculations deduced from (6.16) and (6.8) read:

$$|\Psi\rangle = \sum_{\alpha,\beta} \varphi_{\alpha\beta} |\alpha\beta\rangle = i\psi \sum_{\alpha\beta,i=1}^{3} d_i (\sigma_i \sigma_2)_{\alpha\beta} |\alpha\beta\rangle ,$$

$$\boxed{|\Psi\rangle = i\psi \sum_{\alpha\beta} \langle \beta | (\mathbf{d} \cdot \boldsymbol{\sigma}) \sigma_2 | \alpha \rangle |\alpha\beta\rangle} , \qquad (6.18)$$

$$\psi(\mathbf{d} \cdot \boldsymbol{\sigma}) = -i(\varphi).\sigma_2 \quad \Longrightarrow \quad \boxed{\mathbf{d} = \frac{-i}{2\psi} tr\left((\varphi)(\sigma_2\boldsymbol{\sigma})\right)} = -\frac{i}{2\psi} \sum_{\alpha\beta} (\sigma_2\boldsymbol{\sigma})_{\alpha,\beta} \, \varphi_{\alpha,\beta} ,$$

$$\boxed{\mathbf{d} = \frac{1}{2\psi}[-\Delta^\uparrow(\hat{\mathbf{n}})(\hat{\mathbf{k}}_x + i\hat{\mathbf{k}}_y) + \Delta^\downarrow(\hat{\mathbf{n}})(\hat{\mathbf{k}}_x - i\hat{\mathbf{k}}_y) + 2\Delta^0\hat{\mathbf{k}}_z]} . \qquad (6.19)$$

6.5 Behaviour under Rotations

6.5.1 Rotation in Spin Space

For $\mathbf{d}$ to be a true vector, it should behave appropriately under rotation. $\mathbf{d}$ is representing an order parameter which has both orbital and spin degrees of freedom, but the specificity of odd-parity pairing, leading to the necessity of such a vector representation, is coming from the spin degree of freedom. With the relationship to the Cayley–Klein representation, one should expect that this choice leads to a relationship between rotation in spin space and rotation of $\mathbf{d}$. In fact, the effect of rotations can be calculated both directly and with the generator of rotations. Let's do both methods.

For the direct evaluation, the important point is that the rotation acts simultaneously on both spins. Starting from the expression (6.18) to evaluate the effect of the rotation on the spin part of the order parameter, we get

$$
\begin{aligned}
\mathcal{R}_{\boldsymbol{\Omega}}|\Psi\rangle &= i\psi \sum_{\alpha\beta} \langle\beta|(\mathbf{d}\cdot\boldsymbol{\sigma})\sigma_2|\alpha\rangle \mathcal{R}_{1,\boldsymbol{\Omega}} \otimes \mathcal{R}_{2,\boldsymbol{\Omega}}|\alpha\beta\rangle \\
&= i\psi \sum_{\alpha\beta\gamma\delta} \langle\delta|\mathcal{R}_{\boldsymbol{\Omega}}|\beta\rangle\langle\beta|(\mathbf{d}\cdot\boldsymbol{\sigma})\sigma_2|\alpha\rangle\langle\gamma|\mathcal{R}_{\boldsymbol{\Omega}}|\alpha\rangle|\gamma\delta\rangle \\
&= i\psi \sum_{\alpha\gamma\delta} \langle\delta|\mathcal{R}_{\boldsymbol{\Omega}}(\mathbf{d}\cdot\boldsymbol{\sigma})\sigma_2|\alpha\rangle\langle\gamma|\sigma_2^2\cdot\mathcal{R}_{\boldsymbol{\Omega}}|\alpha\rangle|\gamma\delta\rangle \ .
\end{aligned}
$$

We have

$$
\begin{aligned}
\langle\gamma|\sigma_2^2\cdot\mathcal{R}_{\boldsymbol{\Omega}}|\alpha\rangle &= \sum_{\eta}\langle\gamma|\sigma_2|\eta\rangle\langle\eta|\cos(\Omega/2)\sigma_2 - i\sin(\Omega/2)\sigma_2\hat{\boldsymbol{\Omega}}.\sigma|\alpha\rangle \\
&= \sum_{\eta}(-\langle\eta|\sigma_2|\gamma\rangle)\langle\alpha| - \cos(\Omega/2)\sigma_2 - i\sin(\Omega/2)\sigma_2\hat{\boldsymbol{\Omega}}.\sigma|\eta\rangle \\
&= \langle\alpha|\sigma_2\mathcal{R}_{-\boldsymbol{\Omega}}\sigma_2|\gamma\rangle \ ,
\end{aligned}
$$

as σ_2 is antisymmetric and $\sigma_2(\hat{\boldsymbol{\Omega}}.\sigma)$ is symmetric. So

$$
\begin{aligned}
\mathcal{R}_{\boldsymbol{\Omega}}|\Psi\rangle &= i\psi \sum_{\alpha\gamma\delta} \langle\delta|\mathcal{R}_{\boldsymbol{\Omega}}(\mathbf{d}\cdot\boldsymbol{\sigma})\sigma_2|\alpha\rangle\langle\alpha|\sigma_2\mathcal{R}_{-\boldsymbol{\Omega}}\sigma_2|\gamma\rangle|\gamma\delta\rangle \\
&= i\psi \sum_{\gamma\delta} \langle\delta|\mathcal{R}_{\boldsymbol{\Omega}}(\mathbf{d}\cdot\boldsymbol{\sigma})\mathcal{R}_{-\boldsymbol{\Omega}}\sigma_2|\gamma\rangle|\gamma\delta\rangle \\
&= i\psi \sum_{\gamma\delta} \langle\delta|(\mathcal{R}(\mathbf{d})\cdot\boldsymbol{\sigma})\sigma_2|\gamma\rangle|\gamma\delta\rangle \ .
\end{aligned}
$$

Using (6.10)

$$\mathcal{R}_{\boldsymbol{\Omega}}|\Psi\rangle = |\Psi\left(\mathcal{R}(\mathbf{d})\right)\rangle .$$

So indeed, the effect of a change of the spin quantization axis on the order parameter can be evaluated directly by the corresponding rotation of the (complex) $\mathbf{d}$-vector in 3D. And the calculation above makes a direct connection between the Cayley–Klein transformation and the rather involved definition of the $\mathbf{d}$-vector.

It is also useful (and simple) to evaluate the effect of a rotation using the generator of rotations in spin space: this generator is simply $-\frac{i}{\hbar}\hat{\mathbf{n}}\cdot\mathbf{S}$, where the total spin $\mathbf{S} = \mathbf{S}_1 \otimes \mathbb{1} + \mathbb{1} \otimes \mathbf{S}_2$.

The effect of any operator $\mathbf{O} = O_1 \otimes \mathbb{1} + \mathbb{1} \otimes O_2$ acting in the spin space can be calculated as (remembering that $(\mathbf{d}\cdot\boldsymbol{\sigma})\sigma_2$ is a symmetric matrix and σ_2 an antisymmetric matrix)

$$
\begin{aligned}
\mathbf{O}|\Psi\rangle &= i\psi \sum_{\alpha\beta} \langle\beta|(\mathbf{d}\cdot\boldsymbol{\sigma})\sigma_2|\alpha\rangle \mathbf{O}|\alpha\beta\rangle \\
&= i\psi \sum_{\alpha\beta\gamma} \langle\beta|(\mathbf{d}\cdot\boldsymbol{\sigma})\sigma_2|\alpha\rangle \left(\langle\gamma|O|\alpha\rangle|\gamma\beta\rangle + \langle\gamma|O|\beta\rangle|\alpha\gamma\rangle\right) \\
&= i\psi \sum_{\alpha\beta} \langle\beta|O(\mathbf{d}\cdot\boldsymbol{\sigma})\sigma_2|\alpha\rangle \left(|\alpha\beta\rangle + |\beta\alpha\rangle\right) \quad \text{or} \\
&= i\psi \sum_{\alpha\beta} \left(\langle\beta|O(\mathbf{d}\cdot\boldsymbol{\sigma})\sigma_2|\alpha\rangle + \langle\alpha|O(\mathbf{d}\cdot\boldsymbol{\sigma})\sigma_2|\beta\rangle\right) |\alpha\beta\rangle .
\end{aligned}
\tag{6.20}
$$

Applying (6.20) to the action of generator of rotations in spin space, namely, $-\frac{i}{\hbar}\hat{\mathbf{n}}\cdot\mathbf{S}$, we get

$$
\begin{aligned}
-\frac{i}{\hbar}\hat{\mathbf{n}}.\mathbf{S}|\Psi\rangle &= \psi\frac{1}{2}\sum_{\alpha\beta} \langle\beta|(\hat{\mathbf{n}}\cdot\boldsymbol{\sigma})(\mathbf{d}\cdot\boldsymbol{\sigma})\sigma_2|\alpha\rangle \left(|\alpha\beta\rangle + |\beta\alpha\rangle\right) \\
&= \psi\frac{1}{2}\sum_{\alpha\beta} \langle\beta|(\hat{\mathbf{n}}{\cdot}\mathbf{d})\sigma_2 + i(\hat{\mathbf{n}}\wedge\mathbf{d})\cdot\boldsymbol{\sigma}\sigma_2|\alpha\rangle \left(|\alpha\beta\rangle + |\beta\alpha\rangle\right) .
\end{aligned}
$$

So

$$
-\frac{i}{\hbar}\hat{\mathbf{n}}.S|\Psi\rangle = |\Psi(\hat{\mathbf{n}}\wedge d)\rangle .
\tag{6.21}
$$

So that applying a rotation in spin space amounts to the same rotation of the $\mathbf{d}$-vector [see (6.11)] for an elemental rotation of $\mathbf{d}$: $\mathcal{R}_{\boldsymbol{\Omega},s}|\Psi\rangle = |\Psi\left(\mathcal{R}_{\boldsymbol{\Omega}}(\mathbf{d})\right)\rangle$ (see [3]).

6.5.2   Rotation in Real Space

For rotations in real space (on the orbital degrees of freedom), we should calculate the effect of $-\frac{i}{\hbar}\hat{\mathbf{n}}.L$, with $L = \mathbf{r}\wedge\left(\frac{\hbar}{i}\nabla_\mathbf{r}\right)$. Writing $D(\mathbf{k}) = [\mathbf{d}(\mathbf{k})\cdot\boldsymbol{\sigma}]\sigma_2$, we get

$$-\frac{i}{\hbar}\mathbf{L}|\Psi\rangle = -(\mathbf{r}\wedge\nabla_{\mathbf{r}})\int d\mathbf{k}\,e^{-i\mathbf{k}\cdot\mathbf{r}}\psi\sum_{\alpha\beta}\langle\beta|D(\mathbf{k})|\alpha\rangle|\alpha\beta\rangle$$

$$= \psi\sum_{\alpha\beta}\int d\mathbf{k}\,i(\mathbf{r}\wedge\mathbf{k})e^{-i\mathbf{k}\cdot\mathbf{r}}\langle\beta|D(\mathbf{k})|\alpha\rangle|\alpha\beta\rangle$$

$$= \psi\sum_{\alpha\beta}\int d\mathbf{k}\,\langle\beta|D(\mathbf{k})|\alpha\rangle(\mathbf{k}\wedge\nabla_{\mathbf{k}})e^{-i\mathbf{k}\cdot\mathbf{r}}|\alpha\beta\rangle$$

$$= -\psi\sum_{\alpha\beta,i}\int d\mathbf{k}\,\frac{\partial}{\partial k_i}((\langle\beta|D(\mathbf{k})|\alpha\rangle\mathbf{k})\wedge\hat{\mathbf{k}}_i\,e^{-i\mathbf{k}\cdot\mathbf{r}}|\alpha\beta\rangle$$

$$= -\psi\sum_{\alpha\beta}\int d\mathbf{k}\,e^{-i\mathbf{k}\cdot\mathbf{r}}\langle\beta|(\mathbf{k}\wedge\nabla)D(\mathbf{k})|\alpha\rangle|\alpha\beta\rangle\;,$$

so

$$-\frac{i}{\hbar}\hat{\mathbf{n}}.\mathbf{L}|\Psi\rangle = |\Psi(-i\hat{\mathbf{n}}\cdot\mathbf{L_k}\mathbf{d}(\mathbf{k}))\rangle,\;\text{with}\;\mathbf{L_k} = \mathbf{k}\wedge\frac{1}{i}\nabla_{\mathbf{k}}\;. \tag{6.22}$$

This last expression shows that a rotation in real space acts, as it should, on the order parameter according to its orbital state: p-wave, f-wave, ... for a triplet superconductor, transposed as usual in the reciprocal space.

6.5.3 Change of Quantization Axis: Application to ESP States

In order to get more familiar with rotations of the **d**-vector, let us start with an exercise:

Exercise 1 Consider the very first example of Sect. 6.3.1 to observe the fate of the **d**-vector under a change of orientation of the quantization axis on a simple $|\uparrow\uparrow\rangle$ state. Solution in Sect. 6.11.

Beyond this 'trivial' example, understanding the behaviour under rotation of the **d**-vector is particularly useful to get a more precise idea about some specific spin states. For example, we can easily understand that any state $|\Psi\rangle = \Delta_0|\uparrow\downarrow + \downarrow\uparrow\rangle$ can be considered as an 'equal spin pairing' (ESP) state, with equal weight on $|\uparrow\uparrow\rangle$ and $|\downarrow\downarrow\rangle$ spin components. Indeed, its **d**-vector is simply

$$\mathbf{d} = \frac{1}{\psi}\begin{pmatrix} 0 \\ 0 \\ \Delta_0 \end{pmatrix}\;.$$

Let us rotate the quantization axis by $-\pi/2$ around an axis $\hat{\Omega}$ in the x-y-plane with an angle φ from the x-axis. To get the coordinates of **d** in the new frame, we should rotate it by $\pi/2$ around $\hat{\Omega}$ [remember (6.11)]

$$\hat{\mathbf{\Omega}} = \begin{pmatrix} \cos\varphi \\ \sin\varphi \\ 0 \end{pmatrix} ,$$

$$\mathcal{R}_{\mathbf{\Omega}}(\mathbf{d}) = \sin(\pi/2)\,\hat{\mathbf{\Omega}} \wedge d = \frac{1}{\psi} \begin{pmatrix} \sin\varphi\Delta_0 \\ -\cos\varphi\Delta_0 \\ 0 \end{pmatrix} ,$$

$$\Delta_\uparrow = \psi(-d_x + id_y) = -ie^{i\varphi}\Delta_0 \quad ; \quad \Delta_\downarrow = \psi(d_x + id_y) = -ie^{-i\varphi}\Delta_0 ,$$

which is indeed an ESP state with only ↑↑ and ↓↓ spin components. It is a good exercise to check that, reciprocally, any ESP state with equal weight for the up- and down-spin components can also be written as a pure $S_z = 0$ state for some choice of the quantization axis.

Exercise 6.2 Show that any ESP state with equal weight for the up- and down-spin component can also be written as a pure $|S_z = 0\rangle$ state. Solution in Sect. 6.11.

6.6   Some Uses of the d-Vector Representation

6.6.1   *Amplitude of the d-Vector*

As promised, let us say a few words on the question of normalization of the **d**-vector. For s-wave superconductors, in the simplest cases, we know that the order parameter can be taken as proportional to the gap. Of course, this is wrong in the general case, e.g. gapless superconductivity exists (induced by a critical amount of magnetic impurities for example). But the idea is that $|\psi|^2$ somehow represents the superfluid density. For a spin-triplet superconductor, we can define this quantity as

$$\langle\Psi|\Psi\rangle = \oint \frac{d\Omega}{4\pi} \sum_{\alpha,\beta} \varphi^*_{\beta\alpha}\varphi_{\alpha\beta} = \oint \frac{d\Omega}{4\pi} tr(\varphi^*(\hat{\mathbf{n}})\varphi(\hat{\mathbf{n}}))$$

$$= |\psi|^2 \oint \frac{d\Omega}{4\pi} tr(\sigma_2(\mathbf{d}^* \cdot \boldsymbol{\sigma})(\mathbf{d} \cdot \boldsymbol{\sigma})\sigma_2) = 2|\psi|^2 \oint \frac{d\Omega}{4\pi}|\mathbf{d}(\hat{\mathbf{n}})|^2 .$$

Note that the definition above is coherent with the fact that from the very beginning, we did not normalize (by $\frac{1}{\sqrt{2}}$) the $|\uparrow\uparrow + \downarrow\downarrow\rangle$ component of $|\psi\rangle$ in (6.3). $|\mathbf{d}(\hat{\mathbf{n}})|^2$ can be interpreted as the angular-dependent superconducting (or superfluid) density, and by convention, one takes

$$\oint \frac{d\Omega}{4\pi}|\mathbf{d}(\hat{\mathbf{n}})|^2 = 1 . \tag{6.23}$$

So on calculating averaged quantities $\langle\Psi|O|\Psi\rangle/\langle\Psi|\Psi\rangle$, one should remember that $\langle\Psi|\Psi\rangle = 2|\psi|^2$.

6.6.2 Spin Direction

Up to now, we discussed the properties of the **d**-vector under rotation but we did not unveil the signification of its direction. As announced, it cannot be straightforward, as in the general case, **d** is a complex vector. But it should be related to the spin. So let us calculate $\mathbf{S}|\psi\rangle$, in the same way we performed the calculation of the effects of rotations in spin space, using the generator [Sect. 6.5.1, (6.21)]

$$\mathbf{S}|\Psi\rangle = \frac{i\hbar}{2} \sum_{\alpha\beta} \psi \langle\beta|\sigma(\mathbf{d}\cdot\sigma)\sigma_2|\alpha\rangle\,(|\alpha\beta\rangle + |\beta\alpha\rangle)$$

from which we deduce immediately that

$$\mathbf{d}\cdot\mathbf{S}|\Psi\rangle = 0 . \tag{6.24}$$

This means that if **d** is real (up to a phase factor), it is perpendicular to the direction of the Cooper pairs spin (quantization axis). More explicitly the average spin at a given wave vector **k** of the Fermi surface can be calculated as

$$\frac{\langle\Psi|\mathbf{S}|\Psi\rangle}{\langle\Psi|\Psi\rangle} = \frac{i\hbar}{2} \sum_{\gamma,\delta}\sum_{\alpha,\beta} \langle\gamma|\sigma_2(\mathbf{d}^*\cdot\sigma)|\delta\rangle\langle\beta|(\mathbf{d}\wedge\sigma)\sigma_2|\alpha\rangle\langle\gamma\delta|\,(|\alpha\beta\rangle + |\beta\alpha\rangle)$$

$$= \frac{i\hbar}{2}\, tr\,((\mathbf{d}^*\cdot\sigma)(\mathbf{d}\wedge\sigma)) = \frac{i\hbar}{2}\, tr\,((\mathbf{d}\wedge(\mathbf{d}^*\cdot\sigma)\sigma))$$

$$= \frac{i\hbar}{2}\, (tr\,((\mathbf{d}\wedge\mathbf{d}^*)\mathbb{1}) - i\, tr\,(\mathbf{d}\wedge(\mathbf{d}^*\wedge\sigma)))$$

$$= i\hbar(\mathbf{d}\wedge\mathbf{d}^*) - \frac{\hbar}{2}\, \cancel{tr\,((\mathbf{d}.\sigma)\mathbf{d}^* - \mathbf{d}(\mathbf{d}^*\cdot\sigma))}$$

$$\boxed{\langle\mathbf{S}(\mathbf{k})\rangle = i\hbar\mathbf{d}(\mathbf{k})\wedge\mathbf{d}^*(\mathbf{k})} .$$

$$\tag{6.25}$$

6.6.3 Non-unitary States

The above equation is important. Indeed, if $\mathbf{d}(\mathbf{k})\wedge\mathbf{d}^*(\mathbf{k})$ is non-zero, the state is called a 'non-unitary state' and it has some more involved properties. Moreover, in general, the fact that $\mathbf{d}(\mathbf{k})\wedge\mathbf{d}^*(\mathbf{k})$ is non-zero means that locally, on the Fermi surface, the Cooper pairs spin is non-zero. But it does not mean that globally, the superconductor is spin-polarized. Conversely, if the superconductor is globally spin-polarized, it is necessarily in a non-unitary state, where $\mathbf{d}^*$ is not proportional to $\mathbf{d}$, see Sect. 6.9 on ferromagnetic superconductors.

This notion of 'non-unitary' state is usually bewildering, and it is useful to make some simple calculations in order to get more used to it. For example, we can check what are the conditions under which an ESP state can also be non-unitary. An ESP state has only $\Delta_\uparrow$ and $\Delta_\downarrow$ components, so that its $\mathbf{d}$-vector will be of the form

$$
\mathbf{d} = \psi \begin{pmatrix} \frac{1}{2}(-\Delta^\uparrow + \Delta^\downarrow) \\ -\frac{i}{2}(\ \Delta^\uparrow + \Delta^\downarrow) \\ 0 \end{pmatrix} . \tag{6.26}
$$

Then

$$
\begin{aligned}
\mathbf{d} \wedge \mathbf{d}^* &= (d_x d_y^* - d_y d_x^*)\mathbf{e}_z \\
&= \frac{i|\psi|^2}{4}[(-\Delta^\uparrow + \Delta^\downarrow)(\Delta^{\uparrow*} + \Delta^{\downarrow*}) + (\Delta^\uparrow + \Delta^\downarrow)(-\Delta^{\uparrow*} + \Delta^{\downarrow*})]\mathbf{e}_z \\
&= \frac{-i|\psi|^2}{2}[|\Delta^\uparrow|^2 - |\Delta^\downarrow|^2]\mathbf{e}_z .
\end{aligned}
$$

From (6.23), we get that $\frac{|\psi|^2}{2}(|\Delta^\uparrow|^2 + |\Delta^\downarrow|^2) = 1$. So

$$
i\hbar\mathbf{d} \wedge \mathbf{d}^* = \hbar\frac{[|\Delta^\uparrow|^2 - |\Delta^\downarrow|^2]}{|\Delta^\uparrow|^2 + |\Delta^\downarrow|^2}\mathbf{e}_z . \tag{6.27}
$$

The conclusion is simple: an ESP state is non-unitary only if the amplitude of the $\Delta_\uparrow$ and $\Delta_\downarrow$ components is different on some part of the Fermi surface.

6.6.4   Orbital Moment

In the same way, from (6.22), we calculate that the average orbital moment per Cooper pair is [1]

$$
\begin{aligned}
\frac{\langle\Psi|\mathbf{L}|\Psi\rangle}{\langle\Psi|\Psi\rangle} &= \oint \frac{d\Omega}{4\pi} \frac{\hbar}{2i} \sum_{\alpha,\beta,\gamma,\delta} \langle\gamma|\sigma_2(\mathbf{d}^* \cdot \boldsymbol{\sigma})|\delta\rangle \ \langle\beta|(\mathbf{k} \wedge \nabla_\mathbf{k})(\mathbf{d}(\mathbf{k}) \cdot \boldsymbol{\sigma})\sigma_2|\alpha\rangle\langle\gamma\delta|\alpha\beta\rangle \\
&= \oint \frac{d\Omega}{4\pi} \frac{\hbar}{2i} \ tr\left((\mathbf{d}^* \cdot \boldsymbol{\sigma})(\mathbf{k} \wedge \nabla_\mathbf{k})(\mathbf{d}(\mathbf{k}) \cdot \boldsymbol{\sigma})\right) ,
\end{aligned}
$$

$$
\boxed{\langle\mathbf{L}(\mathbf{k})\rangle = \frac{\hbar}{i} \oint \frac{d\Omega}{4\pi} \sum_i d_i^*(\mathbf{k} \wedge \nabla_\mathbf{k})d_i(k)} . \tag{6.28}
$$

Note that if $\mathbf{d}$ is real, $\langle \mathbf{L} \rangle$ is zero (if not, it would be imaginary!). This appears in (6.28) from

$$\oint \frac{d\Omega}{4\pi} \sum_i d_i^*(\mathbf{k} \wedge \nabla_\mathbf{k}) d_i(k) = \oint \frac{d\Omega}{8\pi} \sum_{i,j} (\mathbf{k} \wedge \mathbf{e}_j) \frac{\partial}{\partial k_j} d_i^2(k)$$

$$= - \oint \frac{d\Omega}{8\pi} \sum_{i,j} (\mathbf{e}_j \wedge \mathbf{e}_j) d_i^2(k) = 0 .$$

A last remark on this point: superconductors for which $\langle \mathbf{L} \rangle$ is non-zero are nowadays called 'chiral superconductors' and quite looked-after for their potential topological properties [4]. Note, however, that if only triplet superconductors can have a non-zero $\langle \mathbf{S} \rangle$, this is not the case for $\langle \mathbf{L} \rangle$: both spin-singlet and spin-triplet can be chiral. Naturally, in case of spin-singlet, the superconductor needs to be unconventional (not s-wave), and intrinsically complex, so that $\langle \mathbf{L} \rangle$ can be non-zero. This is the case, for example, of d-wave superconductors of type "$d + id$" or $(k_x \pm ik_y)k_z \ldots$

6.6.5 Excitation Energy of Quasiparticles

We will not derive the energy spectrum from microscopic theory, just report the results (see [2] for example): for triplet superconductors in a unitary state, the expression of the energy of elementary excitations is very similar to that of singlet anisotropic superconductors, with the $\mathbf{k}$ dependence of the energy gap controlled by the amplitude of $\mathbf{d}(\mathbf{k})$

$$E_k = \sqrt{\xi_k^2 + \Delta^2 \left(|\mathbf{d}(\mathbf{k})|^2 \right)} . \tag{6.29}$$

However, for non-unitary states, two branches appear in the spectrum, depending on the spin orientation of the excitations with respect to $\langle \mathbf{S} \rangle$: it is as if they are 'Zeeman split' by $\langle \mathbf{S} \rangle$. So the energy gap is expressed in such a case as

$$E_k = \sqrt{\xi_k^2 + \Delta^2 \left(|\mathbf{d}(\mathbf{k})|^2 \pm |\mathbf{d}(\mathbf{k}) \wedge \mathbf{d}^*(\mathbf{k})| \right)} . \tag{6.30}$$

Hence, this is another true difference with respect to singlet superconductors.

We can also see how both (6.29) and (6.30) read when using not the $\mathbf{d}$-vector notation, but expression like (6.17) for the order parameter. In the unitary case, the gap $\Delta(\mathbf{k})$ would be expressed as

$$\Delta(\mathbf{k}) = \sqrt{\frac{1}{2} \left(|\Delta^\uparrow(\mathbf{k})|^2 + |\Delta^\downarrow(\mathbf{k})|^2 \right) + |\Delta^0(\mathbf{k})|^2} .$$

And in the case of non-unitary states, if we take the 'simple' example of ESP states, using expression (6.27) for $\mathbf{d} \wedge \mathbf{d}^*$ we derive easily that

$$\Delta(\mathbf{k}) = \sqrt{\Delta^2 \left(|\mathbf{d}(\mathbf{k})|^2 \pm |\mathbf{d}(\mathbf{k}) \wedge \mathbf{d}^*(\mathbf{k})|\right)}$$
$$= |\Delta^\uparrow(\mathbf{k})| \quad \text{or} \quad |\Delta^\downarrow(\mathbf{k})| \, . \tag{6.31}$$

This last expression shows concretely why 'non-unitary states' are a distinctive feature of spin-triplet superconductors. It can be also anticipated that this expression will be particularly useful for ferromagnetic superconductors, where band polarization can lead to a large difference between $|\Delta^\uparrow(\mathbf{k})|$ and $|\Delta^\downarrow(\mathbf{k})|$ (see Sect. 6.9). Expression (6.30) gives a general formula for the two gap values of a non-unitary state, even if it is not an ESP state: as will be seen later, UPt$_3$ in its B-phase could produce such a case (see Sect. 6.8.2.3).

6.7   The Spin–Orbit Issue

Before discussing some emblematic examples of p-wave superconductors, let us say a few words concerning the question of spin–orbit coupling. Indeed, when discussed for real materials (except for superfluid ^{3}He), it covers two different aspects which should be distinguished to avoid confusion. The first is the usual spin–orbit coupling at the atomic scale, discussed already in the normal phase as it prevents the spin $\mathbf{S}$ to be a good quantum number. In a solid, symmetries can help to overcome this problem:

- If the system has an inversion centre and time-reversal symmetry, quasiparticles with a given wave vector $\mathbf{k}$ are necessarily degenerate. This allows to define a 'pseudo-spin 1/2' and to build Cooper pairs with this pseudo-spin state: replacing 'spin' by 'pseudo-spin' is all that is required to keep everything else unchanged.
- If the system has an inversion centre, but not necessarily time-reversal symmetry, then, at least, one can distinguish between odd-parity and even-parity states.
- If there is no inversion centre, but time-reversal symmetry, one can still build Cooper pairs; however, there is no such distinction any more between even- or odd-parity states. Abusively, one can say that singlet and triplet pairings are mixed together. Experimentally, large upper critical fields outpassing the paramagnetic limit are commonly found for such systems.

6.7.1   Spin–Orbit and the Superconducting Order Parameter

However, there is also another issue for 'triplet' superconductors with spin–orbit interaction: we are now speaking of spin–orbit coupling between the spin and orbital parts of the Cooper pairs, as done in the case of superfluid ^{3}He: the problem is that the 'atomic-scale' spin–orbit coupling can be very large (see, for example, what happens in the ^{3}He nuclei!), whereas the coupling between the total orbital moment of the Cooper pair (an object of a coherence length scale) and the total spin of the Cooper pairs can be much weaker. And in real solid, it is very difficult to either calculate (predict) or to measure this spin–orbit interaction. This question is very important because it determines the symmetry group which has to be considered for the classification of the different superconducting states. If spin–orbit is weak, the relative orientation of spin and orbit should be decoupled: so, for example, one can imagine that the spin could reorient 'freely' under the action of an external field. If spin–orbit is strong, the orbital state (the gap nodes for example) expected to be pinned on the crystal lattice will prevent such a reorientation of the spin state. Therefore

- If the spin–orbit interaction is weak, the symmetry group considered for the classification of the possible superconducting states will be $G \otimes U(1) \otimes T \otimes \mathcal{R}_S$, where $\mathcal{R}_S$ are the (3D) rotations in spin space, G is the crystal point group, $U(1)$ the gauge symmetry (always broken in the superconducting state) and T the time-reversal symmetry. Due to $\mathcal{R}_S$, **d** should reorient under field to minimize the Zeeman energy.

- If the spin–orbit interaction is strong, the symmetry group considered for the classification of the possible superconducting states will be $G \otimes U(1) \otimes T$, meaning that the **d**-vector is expected to be 'pinned' on the lattice. In such a case, additional spin anisotropy may appear in the superconducting state, possibly detected, for example, by an anisotropy of the Knight-shift reduction below T_{SC}, or by an anisotropic paramagnetic limitation.

For most of the candidate p-wave superconductors, determining what is the best of the two limits for the description of the system remains an open issue (see, for example, the discussion on UPt$_3$ in Sects. 6.8.2.2 and 6.8.2.3).

6.7.2 Anisotropy of the Susceptibility for the Strong Spin–Orbit Case

Experimentally, an important question when analysing the behaviour of a potential triplet superconductor is the Pauli depairing effect and its anisotropy on the upper critical field, or equivalently, the anisotropy of the change of the Knight shift below T_{SC}, both of which depending on the Cooper pairs spin susceptibility. Supposing that we are in the strong spin–orbit limit, the question is to derive from the possible order parameters, in which directions there will be no change of the electronic spin susceptibility between normal and superconducting phases, and in which directions, if any, there will be at least a partial suppression of this spin susceptibility.

As a matter of fact, it is important to realize that even spin-triplet superconductors, whatever the spin–orbit regime, can present a reduction of the susceptibility for all orientations of the magnetic field. We will see below (Sect. 6.8.1) that superfluid ^{3}He realize, in its B-phase, an A_{1u} state for which $\mathbf{d} \propto \mathbf{k}$. This means that, on each point of the Fermi surface, the order parameter is described by a pure $|S_z = 0\rangle$ state if the quantization z-axis is taken along $\mathbf{k}$. Such a $|S_z = 0\rangle$ state is equivalent to the spin-singlet case as regards susceptibility, leading to a vanishing susceptibility. In fact, it can be shown that for such an A_{1u} state, on average, the susceptibility is reduced to two-third of the normal state susceptibility at $T = 0$, as if for a given field direction, one-third of the spins were in the $|S_z = 0\rangle$ state [1].

At the opposite, for an ESP state, where Cooper pairs are formed only with spins of the same direction, we expect no change of the susceptibility for fields along the quantization axis. However, this does not tell us what to expect in the perpendicular directions.

Maybe the easiest way to understand if the spin susceptibility is reduced or not for a given field direction, and whatever the order parameter, is to rewrite the $\mathbf{d}$-vector with the new quantization axis in this field direction, and check in this representation, whether or not the z-component of the $\mathbf{d}$-vector (corresponding to the amplitude of the $|S_z = 0\rangle$ state for that direction) is zero. Changing the quantization axis, and rewriting the $\mathbf{d}$-vector for this new quantization axis, amounts to rotate the reference frame, or rotate in the opposite direction the $\mathbf{d}$-vector in spin space (see Sect. 6.5.1). It is easy to see [see (6.11) for the rotation of the $\mathbf{d}$-vector] that the z-component of the $\mathbf{d}$-vector, when changing the quantization axis for the x- or y-axis, is, respectively, $-d_x$ or d_y. What it means is that, in the case of strong spin–orbit coupling, where the $\mathbf{d}$-vector cannot reorient depending on the field ($\mathbf{H}$) direction:

- The $|S_z = 0\rangle$ component of the order parameter, where z is the field direction, is proportional to the $\mathbf{d}$-vector projection along the field direction [which generalizes (6.24), which had a physical meaning only for real $\mathbf{d}$-vectors]: if $(\mathbf{d}\cdot\mathbf{H})$ is non-zero, there will be at least a partial suppression of the spin susceptibility and so, Pauli depairing for the upper critical field, for fields applied in this direction.
- Whatever the $\mathbf{d}$-vector, there is always at least one direction, where there will be Pauli depairing (otherwise, $\mathbf{d}$ should be the null vector).

Coming back to the question of ESP states, if the phase between the $\Delta^\uparrow$ and $\Delta^\downarrow$ is constant on the Fermi surface and it is a unitary state, on top of the quantization axis, there is another direction (hence a whole plane) for which there is no change of the spin susceptibility (and no Pauli depairing) and a perpendicular direction for which the spin susceptibility is completely suppressed (see Sect. 6.11.5). For example, if $\mathbf{d}$ is of the form (6.26)

$$\mathbf{d} = \psi \begin{pmatrix} \frac{1}{2}(-\Delta^\uparrow + \Delta^\downarrow) \\ -\frac{i}{2}(\ \Delta^\uparrow + \Delta^\downarrow) \\ 0 \end{pmatrix} ,$$

with $\Delta^\uparrow = \Delta^\downarrow$, then there is no Pauli depairing for fields in the x-z-plane and full Pauli depairing (as in the singlet case) for field along the y-axis. If the ESP state is non-unitary ($|\Delta^\uparrow| \neq |\Delta^\downarrow|$ on some part of the Fermi surface), then for sure, there is at least partial Pauli depairing in the two directions perpendicular to the quantization axis and still no Pauli depairing for fields along the quantization axis.

6.8 d-vector Representation of Some Known (or Suspected) p-Wave Superconductors

6.8.1 Phases of Superfluid ^{3}He

^{3}He has been the very first case for p-wave superconductivity (superfluidity, to be precise!), and it is, beyond contest, a true paradigm for this state of matter. The reasons are that, despite its very low superfluid transition temperature ($\approx$ 1 mK on the melting curve), the spin state and most of the superfluid properties could be identified and studied with tremendous precision thanks to nuclear magnetic resonance (NMR): in superfluid ^{3}He, the Cooper pairs spin is the nuclear spin of the ^{3}He atoms, which are directly probed by NMR. Moreover, the system is rotationally invariant, so with the simplest (spherical) possible Fermi surface, and with spherical harmonics as basis for the irreducible representations of the superconducting order parameter. In $\mathbf{k}$-space, for a p-wave state, they read

$$Y_{11}(\hat{\mathbf{k}}) = -\sqrt{\frac{3}{2}}(\hat{\mathbf{k}}_x + i\hat{\mathbf{k}}_y) \; ; \; Y_{1-1}(\hat{\mathbf{k}}) = \sqrt{\frac{3}{2}}(\hat{\mathbf{k}}_x - i\hat{\mathbf{k}}_y) \; ; \; Y_{10}(\hat{\mathbf{k}}) = -\sqrt{3}\hat{\mathbf{k}}_z \; .$$

(6.32)

It can be explicitly checked that [see (6.22)]

$$\hat{\mathbf{k}}_z \cdot L_k Y_{1m} = \hat{\mathbf{k}}_z \cdot (k \wedge \frac{1}{i}\nabla_{\mathbf{k}} Y_{1m}) = m Y_{1m} \; .$$

(6.33)

Another reason for which superfluid ^{3}He is a paradigm of p-wave superconductors is that it presents a rich phase diagram (shown in Fig. 6.3[1]), with three well-identified phases: two (called A and B) in the temperature–pressure plane, and an additional (A_1) phase under magnetic fields, which are key references. A main topic these days is that of the topological properties of some superconductors, which has been addressed in great detail some tens of years ago for superfluid ^{3}He [5]. Let us examine them quickly.

6.8.1.1 B-Phase

The B-phase of superfluid ^{3}He is simply characterized by

$\mathbf{d}(\mathbf{k}) = \hat{\mathbf{k}}$; or explicitely

$$|\Psi(\hat{\mathbf{k}})\rangle = \psi\left((-\hat{k}_x + i\hat{k}_y)|\uparrow\uparrow\rangle + (\hat{k}_x + i\hat{k}_y)|\downarrow\downarrow\rangle + \hat{k}_z(|\uparrow\downarrow\rangle + |\downarrow\uparrow\rangle)\right) \; .$$

(6.34)

---

[1] Adapted from 'Heliums egenskaper vid låga temperaturer', P. Berglund, Kosmos 1988, s. 63 (Courtesy of the Swedish Physical Society).

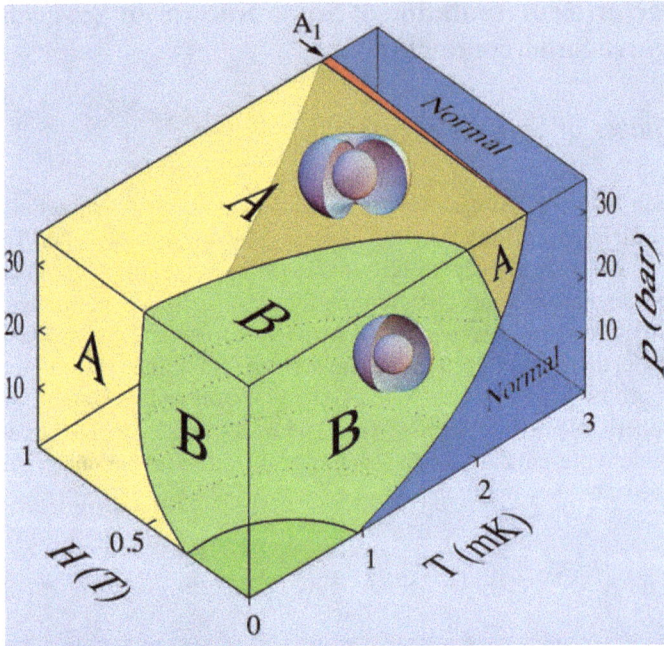

Fig. 6.3 Phase diagram, in the temperature–pressure–magnetic field space, of the superfluid phases of ^{3}He. Three different phases, called *B*, *A* and A_1, corresponding to different symmetries and **d**-vectors have been identified. The gap structure is also shown, represented by the distance between the inner sphere (the Fermi surface) and the exploded view of the outer surface. There is a uniform gap in the *B*-phase, and a nodal gap (with two nodes at the poles) in the *A*-phase. The A_1 phase, which appears only under magnetic field, is non-unitary and the gap is like that of the *A*-phase on the majority spin Fermi sheet, and zero on the other (only half the Fermi surface is paired)

As $\mathbf{d}^*(\mathbf{k}) = \mathbf{d}(\mathbf{k})$, this state is unitary and has both $\langle \mathbf{S} \rangle = 0$ and $\langle \mathbf{L} \rangle = 0$. Moreover, $|\mathbf{d}(\mathbf{k})| = 1$, so that the gap is uniform on the Fermi surface, even though the average of $\mathbf{d}(\mathbf{k})$ is zero.

In the simplest models, this *B*-phase should be the state of lowest free energy, notably due to the fact that the gap is fully open over the whole Fermi surface. However, this state has a reduced spin susceptibility deep in the superfluid state (see the discussion in Sect. 6.7.2). As the pairing mechanism involves spin fluctuations, this may be unfavourable compared to other states, notably ESP states, where such a reduction of the susceptibility is absent (this is the so-called 'feedback' mechanism). Therefore, as seen in Fig. 6.3, another phase, the *A*-phase, is stable notably along the melting line and becomes the dominant phase under field. As will be seen below, this *A*-phase is indeed an ESP state and it has a nodal gap structure.

6.8.1.2 A-Phase

The A-phase of superfluid ^{3}He is simply characterized by

$$\mathbf{d(k)} = \sqrt{\frac{3}{2}}(\hat{k}_y + i\hat{k}_z, 0, 0) \,,$$

$$|\Psi(\hat{\mathbf{k}})\rangle = -\psi\sqrt{\frac{3}{2}}(\hat{k}_y + i\hat{k}_z)(|\uparrow\uparrow\rangle - |\downarrow\downarrow\rangle) \,. \tag{6.35}$$

So, in the A-phase, the excitation gap vanishes for $k_y = k_z = 0$; as shown in Fig. 6.3, it has two nodes on the poles of the Fermi surface. Moreover, this ESP state is unitary, as $\langle \mathbf{S} \rangle = \mathbf{0}$ (since $\mathbf{d^*(k)} \wedge \mathbf{d(k)} = \mathbf{0}$). But $\langle \mathbf{L} \rangle$ is non-zero. In fact, the orbital state is selected (by dipolar coupling), so that $\mathbf{d}$ and $\langle \mathbf{L} \rangle$ are either parallel or antiparallel. Following (6.28)

$$\langle \mathbf{L} \rangle \sim \hbar \quad \frac{}{4\pi} \quad (k_y^2 + k_z^2)\mathbf{e_x} + (\hat{k}_y - i\hat{k}_z)\hat{k}_x(\mathbf{e_z} - i\mathbf{e_y}) \,,$$

$$\langle \mathbf{L} \rangle = \hbar \, \mathbf{e_x} \,. \tag{6.36}$$

The physics of this phase is very rich, notably when considering the weak coupling between the orbital and spin moments due to spin–orbit interaction, the existence of spin currents, the chirality of the excitations close to the nodes... Again, the review by A. J. Leggett [1] is a seminal paper.

6.8.1.3 A_1 Phase

The A_1 phase appears under field with only one spin direction paired: it has the same orbital moment but in addition also a finite average spin. If we keep the same convention for the normalization of $|\Psi(\hat{\mathbf{k}})\rangle$ and $\mathbf{d(k)}$, despite the fact that only half the Fermi surface is paired, we get

$$|\Psi(\hat{\mathbf{k}})\rangle = -\psi\sqrt{3}(\hat{k}_y + i\hat{k}_z)|\uparrow\uparrow\rangle \,,$$

$$\mathbf{d(k)} = \frac{\sqrt{3}}{2}((\hat{k}_y + i\hat{k}_z), -i(\hat{k}_y + i\hat{k}_z), 0) \,. \tag{6.37}$$

For this A_1 phase

$$\langle \mathbf{S} \rangle = i\hbar \oint \frac{d\Omega}{4\pi} (d_x d_y^* - d_y d_x^*) \mathbf{e_z}$$
$$= 2\hbar \oint \frac{d\Omega}{4\pi} |k_y + i\,k_z|^2 \mathbf{e_z}/2 \oint \frac{d\Omega}{4\pi} (k_y^2 + k_z^2) \tag{6.38}$$
$$= \hbar\,\mathbf{e_z}\,,$$
$$\langle \mathbf{L} \rangle = \hbar\,\mathbf{e_z}\,.$$

Its stability arises from the fact that when the Fermi surface is polarized, the density of states increases with k_F, and from the fact that, in ^{3}He, the spin–orbit interaction is very weak. So, the up-spin and down-spin Fermi surfaces are almost completely decoupled and the largest Fermi surface may have a larger transition temperature than the other. Hence, the stability range of this A_1 phase grows under field (see Fig. 6.3).

As we shall see, the situation should be completely different in uranium-based ferromagnetic superconductors, where such a phase is very unlikely due to the coupling between the different Fermi sheets induced by spin–orbit interaction: like for most multigap superconductors, in such a case, there is a unique transition temperature, even if the different gaps may have different sizes. A possible very singular exception will be discussed in Sect. 6.10. Coming back to ^{3}He, the A_1 phase is the paradigm of a non-unitary state, with a finite value of $\langle \mathbf{S} \rangle$, a vanishing gap on one Fermi surface, and a nodal gap (axial gap) identical to that of the A-phase on the other Fermi surface.

6.8.1.4   Planar and Polar Phases

Some other states may also be favoured in ^{3}He, due to peculiar constraints [lower dimensions, aerogel (disordered) background, ...]. These are in any case useful reference states for the more complicated cases of superconductors in crystal lattices. Notably, there is the planar phase and the polar phase, which are derived from the B-phase. The planar phase is defined by

$$\mathbf{d(k)} = \sqrt{\frac{3}{2}}(\hat{k}_x, \hat{k}_y, 0)\,,$$
$$|\Psi(\hat{\mathbf{k}})\rangle = \psi\sqrt{\frac{3}{2}}\left((-\hat{k}_x + i\hat{k}_y)|\uparrow\uparrow\rangle + (\hat{k}_x + i\hat{k}_y)|\downarrow\downarrow\rangle\right)\,. \tag{6.39}$$

Conversely, the polar phase is defined by

$$\mathbf{d(k)} = \sqrt{3}(0, 0, \hat{k}_z)\,,$$
$$|\Psi(\hat{\mathbf{k}})\rangle = \psi\sqrt{3}\hat{k}_z(|\uparrow\downarrow\rangle + |\downarrow\uparrow\rangle)\,. \tag{6.40}$$

These two states have also $\langle \mathbf{S} \rangle = 0$ and $\langle \mathbf{L} \rangle = 0$; however, the planar state has point nodes along the z-axis whereas the polar state has a line of nodes on the equator. Both are also unitary, ESP states (see Sect. 6.5.3).

6.8.2 UPt₃ and Sr₂RuO₄

UPt₃ is a 'heavy fermion' metal, meaning that it is an inter-metallic system with very strong electronic correlation effects, leading to a strong renormalization of the effective mass of the electronic quasiparticles. It has been the first heavy fermion where these effective masses have been directly measured (up to 160 times the free electron mass) on the different Fermi sheets, by quantum oscillations, and it has also been the first superconducting system (after superfluid ³He) where phase transitions between different superconducting phases have been observed (see [6] for a review and Fig. 6.4).

6.8.2.1 Phases of UPt₃

The reasons leading to these phase transitions and the nature of the various superconducting phases have been the subject of many different proposals. There is a global consensus that superconductivity in UPt₃ should be triplet (odd parity). Nevertheless, many questions remain without a definite answer. A first (still open) question, for example, is whether or not the spin component is free to rotate in the hexagonal crystal lattice of UPt₃. This will determine the response of UPt₃ under the application of an external field when it is superconducting. The orbital part (the $\mathbf{k}$-dependence) of the superconducting order parameter is constrained by the broken symmetries in the superconducting state inducing, for example, nodes of the order parameter and so of the gap in some particular directions: if spin–orbit coupling is strong enough, then the $\mathbf{d}$-vector is expected to be pinned in some crystal direction; if spin–orbit coupling is weak, as in superfluid ³He, the $\mathbf{d}$-vector should be free to rotate and the response to a magnetic field should have the same anisotropy as in the normal phase. Because pairing is mainly driven by the $5f$ electrons, spin–orbit coupling is expected to be strong also for the Cooper pair wave function, and pinning of the $\mathbf{d}$-vector is likely. However, this hypothesis has no definite experimental support (see Sect. 6.8.2.3).

6.8.2.2 E₂ᵤ Representation

Among the models assuming such a strong spin–orbit coupling pinning the $\mathbf{d}$-vector in a fixed crystallographic direction, the so-called E_{2u} representation has been strongly developed. It is an 'f-wave' order parameter, which can have various symmetries (six basis functions are necessary to describe the most general order parameter). Among these, the most successful [7] proposes a $\mathbf{d}$-vector with some

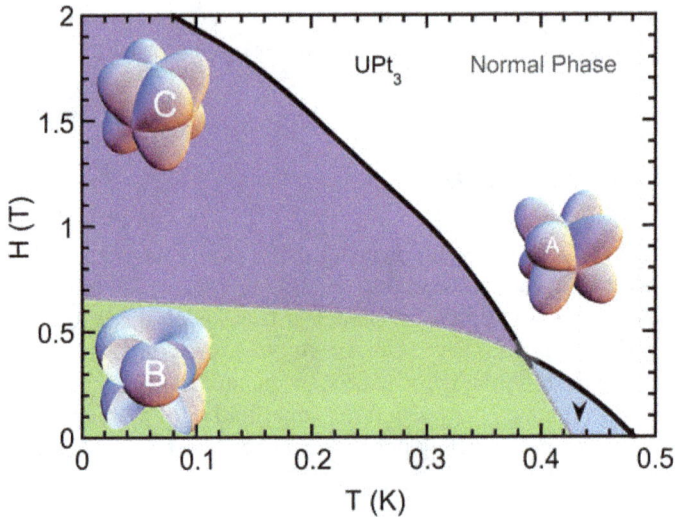

Fig. 6.4 Phase diagram, in the temperature–magnetic field space, of the superconducting phases of UPt₃. Three different phases called A, B and C corresponding to different symmetries and **d**-vectors have been identified. The gap structure is also shown for the different phases, like for superfluid ^{3}He in Fig. 6.3, as proposed for the 'E_{2u} model'. This E_{2u} model is coherent with results from thermal transport and upper critical field measurements, but not with NMR measurements of the Knight shift (see Sect. 6.8.2.3)

accidental restrictions (like the fact that the **d**-vector would only have components along the hexagonal c-axis); with these restrictions, it matches numerous experimental probes.

$$\mathbf{d(k)} = \left(\varphi_A(T)2k_xk_yk_z + \varphi_C(T)k_z(k_x^2 - k_y^2)\right)\mathbf{e}_z . \tag{6.41}$$

In the A-phase, $\varphi_C(T) = 0$, in the C-phase, $\varphi_A(T) = 0$, and in the B-phase, at low temperature and low field, $\varphi_A(T) = \pm i\ \varphi_C(T)$: $\mathbf{d(k)} = \varphi(T)k_z(k_x \pm ik_y)^2\mathbf{e}_z$. For this model, all phases are unitary, but the B-phase is chiral, with a non-zero $\langle\mathbf{L}\rangle$. Muon experiments or recently polar Kerr effect [8] could have detected such a chiral component.

With its pinned **d**-vector, in the A- and C-phases, the spin component is zero along the c-axis ($\mathbf{d} \parallel \mathbf{e}_z$), and for any field direction in the basal plane, the order parameter behaves as an ESP state (see Sect. 6.5.3). So the Pauli spin susceptibility should be suppressed (like for a singlet superconductor), whereas it will be unchanged in the basal plane. This feature, which guided the choice of this E_{2u} representation, can explain the famous 'crossing' of the upper critical fields (H_{c2}) of UPt₃[9] along the basal plane (no Pauli limitation of H_{c2}) and the c-axis (Pauli limitation of H_{c2}).

6.8.2.3 E_{1u} Representation

However, NMR experiments seem to be in contradiction with this interpretation of the crossing of the upper critical fields of UPt$_3$. Indeed Knight-shift measurements, which are the closest to a measure of the spin susceptibility in the superconducting state, found no change in the superconducting state for fields applied along the **a** or **b** directions in the C-phase, but also no change for field along the c-axis except at very low field deep inside the B-phase [11]. Therefore, other models have been proposed for UPt$_3$, which are much closer to the situation of superfluid ^{3}He, with a weak spin–orbit coupling allowing for a field reorientation of the **d**-vector as long as the field is 'strong enough'. Knight-shift measurements can tell nothing on the gap nodes, but combining angle-dependent thermal conductivity measurements [10] with the NMR result, a E_{1u} scenario has been proposed, predicting a non-chiral state [$\mathbf{d}(\mathbf{k}) \propto (5\hat{k}_c^2 - 1)(\hat{k}_a \mathbf{e}_b + \hat{k}_b \mathbf{e}_c)$ in the B-phase]. This scenario can also more easily give account of some other features of the phase diagram (like the existence of a tetracritical point in all field directions): see Fig. 6.5. For the different phases, in this model, the **d**-vector would be as shown in Table 6.1.

So, as can be seen from Table 6.1, at low fields the **d**-vector does not depend on the field orientation [check the A-phase and B-phase-(low H) lines of the table]. In the B-phase, for **H** ∥ **c**, there is a field-induced reorientation of the **d**-vector: at low field, with $\mathbf{d} \propto (\hat{k}_a \mathbf{e}_b + \hat{k}_b \mathbf{e}_c)$, the **d**-vector is not perpendicular to the c-axis (except on the line $\hat{k}_b = 0$) so there is a finite $|S_z = 0\rangle$ component of the spin along the field. This would be imposed by the orbital part of the wave function and spin–orbit interaction or coupling to the small antiferromagnetic moments acting as a symmetry breaking field. But for fields above 0.22 T, with $\mathbf{d} \propto (\hat{k}_a \mathbf{e}_b + \hat{k}_b \mathbf{e}_a)$, the **d**-vector is real and always perpendicular to the c-axis, so we know that it is equivalent to an ESP state in that direction. Hence, the field-induced rotation of the **d**-vector. In the high-field C-phase, where the Pauli limitation could be at play, we note that in Table 6.1 the **d**-vector is always perpendicular to the field direction, so that again, it is an ESP state explaining the observed absence of change of the Knight shift (but in contradiction with the H_{c2} anisotropy). Note also that all these features are preserved if in the B-phase, the **d**-vector is a complex combination of $\mathbf{e}_b$ and $\mathbf{e}_c$, or $\mathbf{e}_b$ and $\mathbf{e}_a$: $(\hat{k}_a \mathbf{e}_b \pm i\hat{k}_b \mathbf{e}_c)$; $(\hat{k}_a \mathbf{e}_b \pm i\hat{k}_b \mathbf{e}_a)$.

Then, the B-phase would be chiral (as in the E_{2u} model), but also non-unitary $(\mathbf{d}(\mathbf{k}) \wedge \mathbf{d}^*(\mathbf{k}) \neq 0$, see Sect. 6.6.3). So was the original proposal in [11]. It is an interesting example of a non-unitary state with no global spin polarization, e.g. with $\mathbf{d} \propto (5\hat{k}_c^2 - 1)(\hat{k}_a \mathbf{e}_b + i\hat{k}_b \mathbf{e}_a)$, we derive from (6.25) and (6.31) that $\langle \mathbf{S}(\mathbf{k})\rangle \propto (5\hat{k}_c^2 - 1)^2\hat{k}_a\hat{k}_b\mathbf{e}_c$ and $\Delta^\uparrow(\mathbf{k}) \propto |(5\hat{k}_c^2 - 1)(\hat{k}_a + \hat{k}_b)|$, $\Delta^\downarrow(\mathbf{k}) \propto |(5\hat{k}_c^2 - 1)(\hat{k}_a - \hat{k}_b)|$. So indeed, averaging over the Fermi surface leads to no net spin and equal averaged gap amplitudes for up-spins and down-spins, even though they are different for most **k** of the Fermi surface.

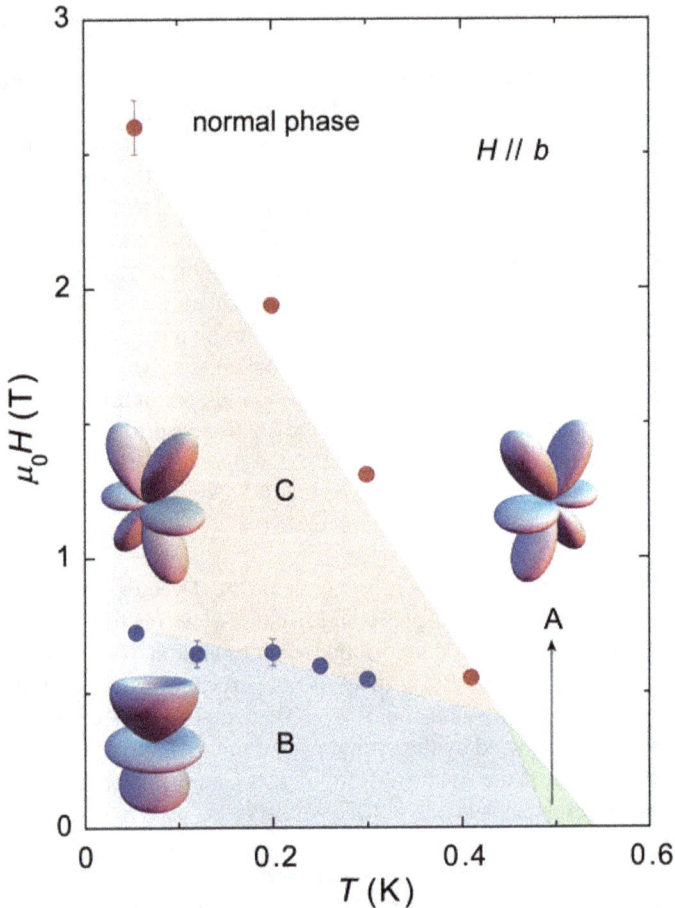

Fig. 6.5 Phase diagram in the temperature–magnetic field space of the superconducting phases of UPt$_3$: the same phases as in Fig. 6.4 are represented, but different gap structures are proposed, according to the E_{1u} symmetry (the Fermi surface is reduced to a point in this representation). The magnetic field H is parallel to the lattice vector **b**.

6.8.2.4   *p*-Wave Superconductivity in Sr$_2$RuO$_4$

This system has also early been proposed as a candidate for *p*-wave superconductivity. In fact, due to the accessible range for the studies (T_{SC} is slightly larger than 1 K), and the absence of radioactive elements, as well as the popularity of oxides, it is certainly the most studied '*p*-wave' superconductor ever (for a review, see [12]). The most 'fashionable' order parameter is very similar to the superfluid ^{3}He A-phase

$$\mathbf{d(k)} \propto (k_x \pm i\, k_y)\mathbf{e}_z \, . \tag{6.42}$$

Table 6.1 The **d**-vector for the various phases and magnetic field orientations in the E_{1u} model of UPt$_3$, as proposed in [10, 11]. The distinction 'low field' or 'high field' is meaningful only in the B-phase for **H** $\parallel$ **c**, and is suggested by NMR measurements which show a decrease of the Knight shift in the whole B- and A-phases for **H** $\parallel$ **b**, and for **H** $\parallel$ **c** in the B-phase only, and for fields lower than 0.22 T. Above this value, the **d**-vector would rotate and the change of the Knight shift disappears (see [11])

Phase	**H** $\parallel$ **a**	**H** $\parallel$ **b**	**H** $\parallel$ **c**
A	$(5\hat{k}_c^2 - 1)(\hat{k}_a\,\mathbf{e}_b)$	$(5\hat{k}_c^2 - 1)(\hat{k}_a\,\mathbf{e}_b)$	$(5\hat{k}_c^2 - 1)(\hat{k}_a\,\mathbf{e}_b)$
C	$(5\hat{k}_c^2 - 1)(\hat{k}_b\,\mathbf{e}_c)$	$(5\hat{k}_c^2 - 1)(\hat{k}_b\,\mathbf{e}_c)$	$(5\hat{k}_c^2 - 1)(\hat{k}_b\,\mathbf{e}_a)$
B (low H)	$(5\hat{k}_c^2 - 1)(\hat{k}_a\,\mathbf{e}_b + \hat{k}_b\,\mathbf{e}_c)$	$(5\hat{k}_c^2 - 1)(\hat{k}_a\,\mathbf{e}_b + \hat{k}_b\,\mathbf{e}_c)$	$(5\hat{k}_c^2 - 1)(\hat{k}_a\,\mathbf{e}_b + \hat{k}_b\,\mathbf{e}_c)$
B (high H)	$(5\hat{k}_c^2 - 1)(\hat{k}_a\,\mathbf{e}_b + \hat{k}_b\,\mathbf{e}_c)$	$(5\hat{k}_c^2 - 1)(\hat{k}_a\,\mathbf{e}_b + \hat{k}_b\,\mathbf{e}_c)$	$(5\hat{k}_c^2 - 1)(\hat{k}_a\,\mathbf{e}_b + \hat{k}_b\,\mathbf{e}_a)$

It is a unitary chiral state with $\langle \mathbf{L} \rangle = \pm\hbar\,\mathbf{e_z}$ and point nodes along the c-axis. However, due to its quasi-2D Fermi surface, there is no **k** vector on the Fermi surface at the node position. There are many contradictory experiments on this system and several have claimed to have detected or refuted the time-reversal symmetry breaking in this compound (this is also true for the B-phase of UPt$_3$). So today, there is still no firm conclusion on whether or not (6.42) is the correct gap symmetry for Sr$_2$RuO$_4$, and even on whether or not it is really a p-wave superconductor. Indeed, the most recent NMR studies corrected previous results and demonstrate now that the Knight shift does decrease in the superconducting phase when the field is applied in the basal plane, ruling out one of the strongest support for an order parameter of the above form [13, 14].

6.9 Ferromagnetic Superconductors

Since 2000, three systems with a true homogeneous coexistence of ferromagnetic order and superconductivity have been discovered; all of them are uranium based. The first, UGe$_2$ [15], is only superconducting under pressure, the other two, URhGe [16] and UCoGe [17], are superconducting at ambient pressure. In the three cases, the same $5f$ electrons from the uranium ions are responsible for the ferromagnetic and the superconducting orders, and the Curie temperature (T_{Curie}) is always larger than the superconducting transition temperature (T_{SC}). Intuitively, these two orders seem antagonistic, as it is known that superconductivity is suppressed by large fields. However, it is important to be more precise in order to understand why and how ferromagnetism and superconductivity might coexist.

The first point to have in mind is the two kinds of magnetic fields associated with ferromagnetic order: there is an internal magnetic field B_{int} also called the 'dipolar field', arising from the spontaneous magnetization in the sample ($\mathbf{B} \approx \mathbf{M}$,

Table 6.2 Orders of magnitude of some important parameters, including an estimate of the internal (dipolar) magnetic field B_{int} (coming from the spontaneous magnetization) and the effective exchange field B_{exc}, in the three known uranium-based ferromagnetic superconductors. For UGe$_2$, which is superconducting only under pressure, we have indicated the Curie temperature and the ordered moment μ_{ord} at the pressure of 1.2 GPa, where T_{SC} is maximum. For B_{exc} we only give a lower bound deduced from the value of the Curie temperature

	UGe$_2$ (1.2 GPa)	URhGe	UCoGe
T_{Curie}	35 K	9.5 K	2.5 K
T_{SC}	0.8 K	0.25 K	0.5 K
μ_{ord}	$\approx 1\ \mu_B$	0.4 μ_B	0.05 μ_B
$B_{int} \approx M$	0.2 T	0.09 T	0.1 T
$B_{exc} > \frac{k_B}{\mu_B} T_{Curie}$	50 T	13 T	4.5 T

if one neglects demagnetization effects) and there is the exchange field B_{exc} which is a very short range effective magnetic field, acting only on the electron spins, and arising from the Coulomb interaction and the exclusion principle. This exchange field appears in a mean-field treatment of the spin–spin exchange interaction term. These two fields have very different orders of magnitude. The first is rather small in these systems, owing to the weak ordered moment (see Table 6.2); indeed, the three compounds, when they are not in the itinerant limit, remain close to it, so that this internal field is in any case much smaller than the (large) orbital upper critical field. However, the exchange field, whose scale is fixed by $k_B T_{Curie}/\mu_B$, is much larger than the Pauli paramagnetic limit (of the order of $2k_B T_{SC}/\mu_B$). Table 6.2 reports the values of these fields for the different compounds; a recent review has been published in [18].

6.9.1   ESP States

In these uranium-based ferromagnetic superconductors, superconductivity sets in below T_{Curie}, so that Cooper pairs are formed from a spin-polarized Fermi surface. Intuitively, one can guess that if the polarization is large enough (typically, if the difference in the Fermi wave vectors is larger than the inverse coherence length), this leaves little choice but to form Cooper pairs 'independently' on the Fermi sheets with different spin orientations (see Fig. 6.6). In other words, the strong exchange field present in the uranium-based ferromagnetic superconductors seems only consistent with an odd-parity (triplet) superconducting order parameter. Moreover, in case of large polarization of the bands (compared to Δ), one expects only ESP states to be favoured. Choosing the quantization axis along the easy axis

$$|\Psi\rangle = \Delta^\uparrow |\uparrow\uparrow\rangle + \Delta^\downarrow |\downarrow\downarrow\rangle . \tag{6.43}$$

Fig. 6.6 Scheme of the spin-dependent density of states in a ferromagnetic metal. If superconductivity develops on spin-polarized Fermi surfaces, due to the difference of wave vectors at the Fermi level, mainly up-up and down-down Cooper pairs can be formed (short arrows). This leads to an ESP state with different weights for the majority and minority spins, so to a non-unitary ESP state

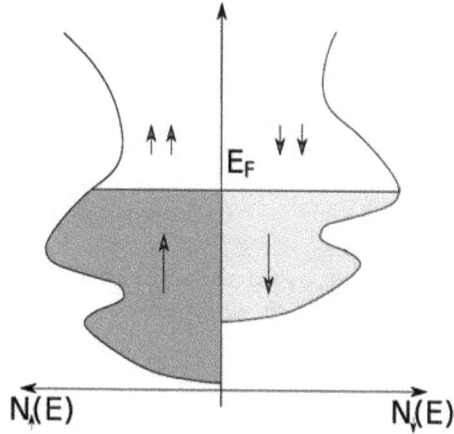

In such a case, the **d**-vector would have the general form:

$$\begin{cases} d_x = \dfrac{1}{2\psi}(-\Delta^\uparrow + \Delta^\downarrow) \\[2mm] d_y = \dfrac{-i}{2\psi}(\Delta^\uparrow + \Delta^\downarrow) \\[2mm] d_z = 0 \end{cases} \tag{6.44}$$

Then $\langle \mathbf{S} \rangle$ at a given $\mathbf{k}$ is

$$\begin{aligned} \langle \mathbf{S} \rangle &= i\,\hbar(\mathbf{d} \wedge \mathbf{d}^*) \\ &= -\frac{\hbar}{4\psi}\left((-\Delta^\uparrow + \Delta^\downarrow)(\Delta^{\uparrow *} + \Delta^{\downarrow *}) + \Delta^\uparrow + \Delta^\downarrow)(-\Delta^{\uparrow *} + \Delta^{\downarrow *})\right)\mathbf{e}_z \\ &= \hbar\frac{\left(|\Delta^\uparrow|^2 - |\Delta^\downarrow|^2\right)}{\left(|\Delta^\uparrow|^2 + |\Delta^\downarrow|^2\right)}\mathbf{e}_z . \end{aligned}$$

$$(6.45)$$

This has been already seen when discussing ESP states [see (6.27)]. It is a very natural result; for an ESP state, there is a finite spin (and non-unitary state) if and only if the weights of the $|\uparrow\uparrow\rangle$ and $|\downarrow\downarrow\rangle$ components are unbalanced. Conversely, for a ferromagnetic superconductor with at least partial band polarization, one does expect to have such a non-unitary state, which should be an ESP for a 'strong enough' exchange field. It could also be chiral, but it should be at least non-unitary. Before that, let us see what are the possible order parameter from group theory considerations for orthorhombic systems [19].

6.9.2 Symmetries

Only two one-dimensional representations are left, called A and B, due to the low orthorhombic symmetry. In the paramagnetic state, the first one (A) looks very much like the B-phase of superfluid ^{3}He

$$\mathbf{d}_A(\mathbf{k}) \propto u_x \hat{k}_x \mathbf{e}_x + u_y \hat{k}_y \mathbf{e}_y + u_z \hat{k}_z \mathbf{e}_z , \tag{6.46}$$

where u_x, u_y, u_z are real functions of $\mathbf{k}$ with full orthorhombic symmetry. So in the paramagnetic state, we start from

$$\begin{cases} \Delta_A^\uparrow = -u_x \hat{k}_x + i u_y \hat{k}_y \\ \Delta_A^\downarrow = +u_x \hat{k}_x + i u_y \hat{k}_y \\ \Delta_A^0 = u_z \hat{k}_z \end{cases} . \tag{6.47}$$

In the ferromagnetic state, the amplitude (and possibly the phase) of the $\Delta_A^\uparrow$ and $\Delta_A^\downarrow$ components can differ. So the order parameter should read

$$\begin{cases} \Delta_A^\uparrow = \eta^\uparrow(-u_x \hat{k}_x + i u_y \hat{k}_y) = -\eta_x^\uparrow \hat{k}_x + i \eta_y^\uparrow \hat{k}_y \\ \Delta_A^\downarrow = \eta^\downarrow(+u_x \hat{k}_x + i u_y \hat{k}_y) = \eta_x^\downarrow \hat{k}_x + i \eta_y^\downarrow \hat{k}_y \\ \Delta_A^0 = \eta^0 u_z \hat{k}_z = \eta_z^0 \hat{k}_z \end{cases} \tag{6.48}$$

with no phase difference between the complex amplitudes $\eta^\uparrow$, $\eta^\downarrow$, η^0. In terms of $\mathbf{d}$-vectors

$$\begin{aligned} \mathbf{d} &= \frac{1}{2\psi}[-\Delta^\uparrow(\hat{\mathbf{n}})(\mathbf{e}_x + i\mathbf{e}_y) + \Delta^\downarrow(\hat{\mathbf{n}})(\mathbf{e}_x - i\mathbf{e}_y)] + \Delta^0 \mathbf{e}_z \\ &= \frac{1}{2\psi}[(\eta_x^\uparrow \hat{k}_x - i\eta_y^\uparrow \hat{k}_y)(\mathbf{e}_x + i\mathbf{e}_y) + (\eta_x^\downarrow \hat{k}_x + i\eta_y^\downarrow \hat{k}_y)(\mathbf{e}_x - i\mathbf{e}_y)] + \eta_z^0 \hat{k}_z \mathbf{e}_z \\ &= \frac{1}{2\psi}\left\{\left[\left(\eta_x^\uparrow + \eta_x^\downarrow\right)\hat{k}_x - i\left(\eta_y^\uparrow - \eta_y^\downarrow\right)\hat{k}_y\right]\mathbf{e}_x \right. \\ &\quad \left. + \left[\left(\eta_y^\uparrow + \eta_y^\downarrow\right)\hat{k}_y + i\left(\eta_x^\uparrow - \eta_x^\downarrow\right)\hat{k}_x\right]\mathbf{e}_y + 2\eta_z^0 \hat{k}_z \mathbf{e}_z\right\} . \end{aligned}$$

$$\tag{6.49}$$

Similarly for the B-phase, the order parameter reads

$$\begin{cases} \Delta_B^\uparrow = \zeta_z^\uparrow \hat{k}_z \\ \Delta_B^\downarrow = \zeta_z^\downarrow \hat{k}_z \\ \Delta_B^0 = \zeta_x^0 \hat{k}_x + i\zeta_y^0 \hat{k}_y \end{cases} \tag{6.50}$$

$$\mathbf{d} = \frac{1}{2\psi}[-\Delta^\uparrow(\hat{\mathbf{n}})(\mathbf{e}_x + i\mathbf{e}_y) + \Delta^\downarrow(\hat{\mathbf{n}})(\mathbf{e}_x - i\mathbf{e}_y)] + \Delta^0\mathbf{e}_z$$

$$= \frac{1}{2\psi}[(-\zeta_z^\uparrow\hat{k}_z)(\mathbf{e}_x + i\mathbf{e}_y) + (\zeta_z^\downarrow\hat{k}_z)(\mathbf{e}_x - i\mathbf{e}_y)] + (\zeta_x^0\hat{k}_x + i\zeta_y^0\hat{k}_y)\mathbf{e}_z \qquad (6.51)$$

$$= \frac{1}{2\psi}\left\{ \left(\zeta_z^\downarrow - \zeta_z^\uparrow\right)\hat{k}_z\mathbf{e}_x - i\left(\zeta_z^\uparrow + \zeta_z^\downarrow\right)\hat{k}_z\mathbf{e}_y + 2\left(\zeta_x^0\hat{k}_x + i\zeta_y^0\hat{k}_y\right)\mathbf{e}_z \right\}.$$

In the general case, neither the A- nor the B-phase have symmetry-enforced nodes. However, if an ESP state is enforced by strong band splitting, meaning that the Δ^0 component vanishes in (6.48) and (6.50), then

- the order parameter of the A-phase vanishes for $k_x = k_y = 0$: the A-phase has poles on the z-axis and
- the order parameter of the B-phase vanishes for $k_z = 0$: the B-phase has a line of nodes on the equator.

This is correct, but maybe more important insights on these ferromagnetic supercon-ductors can be learned, notably on the relationship between up and down components, from a more general microscopic model [19]. The equations look unfriendly at first sight, but at the end, a nice physical picture emerges.

6.9.3 Microscopic Model

For these ferromagnetic superconductors, all models start from the same pairing interaction, supposed to arise from the magnetic interactions: this was already pro-posed in the de Gennes book on superconductivity ([20] page 104)! So they start from a Hamiltonian

$$H_{int} = -\frac{1}{2}\mu_B^2 I^2 \int d^3\mathbf{r}d^3\mathbf{r}' S_i(\mathbf{r})\chi_{ij}(\mathbf{r} - \mathbf{r}')S_j(\mathbf{r}') , \qquad (6.52)$$

where I is an exchange constant and χ_{ij} the medium magnetic susceptibility (matrix). The models differ notably on the expression for this susceptibility. However, starting from such an Hamiltonian, the derived gap equations necessarily couple the different components of the order parameter. In the case of ferromagnetic superconductors, a peculiarity found in all systems is that the susceptibility has a marked uniaxial anisotropy (Ising type), so that if z is the easy magnetization axis, χ_{zz} will be a dominant term in the susceptibility matrix.

In the following, we will just present and comment the linearized gap equations (in the weak-coupling limit) to show where the approximations come into play and what are the physical consequences. To understand the derivation of these equations, please refer to [19]. The gap equations (at T_{SC}) read

$$
\left\{
\begin{aligned}
\Delta^{\uparrow}(\mathbf{k}) &= -T\sum_{n}\sum_{\mathbf{k}'}[V^{\uparrow\uparrow}G^{\uparrow}G^{\uparrow}\Delta^{\uparrow}(\mathbf{k}') \\
&\quad + V^{\uparrow\downarrow}G^{\downarrow}G^{\downarrow}\Delta^{\downarrow}(\mathbf{k}') + V^{\uparrow0}(G^{\downarrow}G^{\uparrow}+G^{\uparrow}G^{\downarrow})\Delta^{0}(\mathbf{k}')] \\
\Delta^{\downarrow}(\mathbf{k}) &= -T\sum_{n}\sum_{\mathbf{k}'}[V^{\downarrow\uparrow}G^{\uparrow}G^{\uparrow}\Delta^{\uparrow}(\mathbf{k}') \\
&\quad + V^{\downarrow\downarrow}G^{\downarrow}G^{\downarrow}\Delta^{\downarrow}(\mathbf{k}') + V^{\downarrow0}(G^{\downarrow}G^{\uparrow}+G^{\uparrow}G^{\downarrow})\Delta^{0}(\mathbf{k}')] \\
\Delta^{0}(\mathbf{k}) &= -T\sum_{n}\sum_{\mathbf{k}'}[V^{0\uparrow}G^{\uparrow}G^{\uparrow}\Delta^{\uparrow}(\mathbf{k}') \\
&\quad + V^{0\downarrow}G^{\downarrow}G^{\downarrow}\Delta^{\downarrow}(\mathbf{k}') + V^{00}(G^{\downarrow}G^{\uparrow}+G^{\uparrow}G^{\downarrow})\Delta^{0}(\mathbf{k}')]
\end{aligned}
\right.
$$

$$(6.53)$$

where the different interaction terms $V^{\alpha\beta}$ are expressed from the susceptibility by

$$
\left\{
\begin{aligned}
V^{\uparrow\uparrow} &= -\mu_B^2 I^2 \chi_{zz}^u & &; \quad V^{\downarrow\downarrow} = -\mu_B^2 I^2 \chi_{zz}^u \\
V^{\uparrow\downarrow} &= -\mu_B^2 I^2(\chi_{xx}^u - \chi_{yy}^u - 2i\chi_{xy}^u)\;;\; & V^{\downarrow\uparrow} &= -\mu_B^2 I^2(\chi_{xx}^u - \chi_{yy}^u + 2i\chi_{xy}^u) \\
V^{\uparrow0} &= -\mu_B^2 I^2(\chi_{xz}^u - i\chi_{yz}^u + 2i\chi_{xy}^u)\;;\; & V^{\downarrow0} &= -\mu_B^2 I^2(-\chi_{xz}^u - i\chi_{yz}^u + 2i\chi_{xy}^u) \cdot \\
V^{00} &= -\mu_B^2 I^2 \dfrac{\chi_{xx}^u + \chi_{yy}^u - \chi_{zz}^u}{2}
\end{aligned}
\right.
$$

$$(6.54)$$

In the above equations,

$$V^{\alpha\beta} = V^{\alpha\beta}(\mathbf{k},\mathbf{k}') \text{ and } \chi_{ij}^u = \chi_{ij}^u(\mathbf{k},\mathbf{k}') = \tfrac{1}{2}\left(\chi_{ij}(\mathbf{k}-\mathbf{k}') - \chi_{ij}(\mathbf{k}+\mathbf{k}')\right).$$

Strong band polarization (like that due to an 'exchange field') much larger than T_{SC} means that we can cancel all terms with Green functions arising from bands of opposite spins ($G^{\downarrow}G^{\uparrow}$-type terms). However, (6.53) shows that this is not enough to ensure an ESP state with $\Delta^0(\mathbf{k}) = 0$. Indeed, the different components are all coupled together as a multigap system and a non-zero $\Delta^0(\mathbf{k})$ can be induced by the $\Delta^{\uparrow}$, $\Delta^{\downarrow}$ components

$$
\left\{
\begin{aligned}
\Delta^{\uparrow}(\mathbf{k}) &= -T\sum_{n}\sum_{\mathbf{k}'}[V^{\uparrow\uparrow}G^{\uparrow}G^{\uparrow}\Delta^{\uparrow}(\mathbf{k}') + V^{\uparrow\downarrow}G^{\downarrow}G^{\downarrow}\Delta^{\downarrow}(\mathbf{k}')] \\
\Delta^{\downarrow}(\mathbf{k}) &= -T\sum_{n}\sum_{\mathbf{k}'}[V^{\downarrow\uparrow}G^{\uparrow}G^{\uparrow}\Delta^{\uparrow}(\mathbf{k}') + V^{\downarrow\downarrow}G^{\downarrow}G^{\downarrow}\Delta^{\downarrow}(\mathbf{k}')] \\
\Delta^{0}(\mathbf{k}) &= -T\sum_{n}\sum_{\mathbf{k}'}[V^{\uparrow0}G^{\uparrow}G^{\uparrow}\Delta^{\uparrow}(\mathbf{k}') + V^{\downarrow0}G^{\downarrow}G^{\downarrow}\Delta^{\downarrow}(\mathbf{k}')]
\end{aligned}
\right.
$$

$$(6.55)$$

To go further, we need to make approximations based on the characteristics of the susceptibility. All non-diagonal terms of χ_{ij} are zero at $k = 0$, but, in principle, can be finite at finite $\mathbf{k}$. In a Landau framework, they would arise from gradient terms of the form $\frac{\partial M_i}{\partial x_j}$, so from spin–orbit coupling. If the spin–orbit coupling is weak enough, we can further neglect these terms, then $V^{\uparrow0}$ and $V^{\downarrow0}$ are suppressed and we do get an ESP state: $\Delta^0(\mathbf{k}) = 0$. However, in uranium-based systems, spin–orbit coupling is usually considered as 'strong' and most models suppose that the **d**-vector has a fixed direction, imposed by the orbital part of the order parameter and ... spin–orbit coupling. Therefore, considering these systems as pure ESP states is probably only an approximation: spin–orbit coupling most likely induces a (small?) Δ^0 finite component, even with strong band polarization! But this is not the only surprise which emerges from these microscopic equations. Even if we suppose that $\Delta^0 = 0$, another counter-intuitive result emerges. The equations in the 'ESP approximation' are written as

$$
\begin{cases}
\Delta^\uparrow(\mathbf{k}) = -T \sum_n \sum_{\mathbf{k}'} [V^{\uparrow\uparrow} G^\uparrow G^\uparrow \Delta^\uparrow(\mathbf{k}') + V^{\uparrow\downarrow} G^\downarrow G^\downarrow \Delta^\downarrow(\mathbf{k}')] \\
\Delta^\downarrow(\mathbf{k}) = -T \sum_n \sum_{\mathbf{k}'} [V^{\downarrow\uparrow} G^\uparrow G^\uparrow \Delta^\uparrow(\mathbf{k}') + V^{\downarrow\downarrow} G^\downarrow G^\downarrow \Delta^\downarrow(\mathbf{k}')]
\end{cases}
\tag{6.56}
$$

with

$$
\begin{cases}
V^{\uparrow\uparrow} = V^{\downarrow\downarrow} = -\mu_B^2 I^2 \chi_{zz}^u \\
V^{\uparrow\downarrow} = V^{\downarrow\uparrow} = -\mu_B^2 I^2 (\chi_{xx}^u - \chi_{yy}^u)
\end{cases}
\tag{6.57}
$$

It corresponds to the equations of a two-band superconductor, with intra-band coupling controlled by the longitudinal susceptibility χ_{zz}^u and inter-band coupling controlled by transverse susceptibilities.

For such ESP states, the possible order parameters of ferromagnetic superconductors are the above-mentioned A or B states, with **d**-vector

$$
\begin{aligned}
\mathbf{d}_A &= \frac{1}{2\psi} \left\{ \left[\left(\eta_x^\uparrow + \eta_x^\downarrow \right) \hat{k}_x - i \left(\eta_y^\uparrow - \eta_y^\downarrow \right) \hat{k}_y \right] \mathbf{e}_x \right. \\
&\quad \left. + \left[\left(\eta_y^\uparrow + \eta_y^\downarrow \right) \hat{k}_y + i \left(\eta_x^\uparrow - \eta_x^\downarrow \right) \hat{k}_x \right] \mathbf{e}_y \right\} , \\
\mathbf{d}_B &= \frac{1}{2\psi} \left\{ \left(\eta_z^\downarrow - \eta_z^\uparrow \right) \hat{k}_z \mathbf{e}_x - i \left(\eta_z^\uparrow + \eta_z^\downarrow \right) \hat{k}_z \mathbf{e}_y \right\} .
\end{aligned}
\tag{6.58}
$$

At the same level of approximation, equations for η_x, η_y are decoupled. So, for both the A and B states, the equation for the largest T_{SC} is that of a two-band superconductor (where ϵ is a characteristic energy)

$$
T_{SC} = \epsilon \exp\left(-\frac{1}{g} \right) ,
$$

$$
g = \frac{g_1^\uparrow + g_1^\downarrow}{2} + \sqrt{\frac{\left(g_1^\uparrow - g_1^\downarrow \right)^2}{4} + g_2^\uparrow g_2^\downarrow} .
\tag{6.59}
$$

Here, $g_1^{\uparrow,\downarrow} \propto V^{\uparrow\uparrow,\downarrow\downarrow} \propto \chi_{zz}$, $g_2^{\uparrow,\downarrow} \propto V^{\uparrow\downarrow,\downarrow\uparrow} \propto (\chi_{xx} - \chi_{yy})$. Equation (6.59) shows that, as for any two-band superconductor, T_{SC} should increase if the inter-band coupling is increased. In this case, T_{SC} should increase if the difference between the transverse susceptibilities increases. This is surprising for two reasons.

The first is that it was believed, since the pioneering work of D. Fay and J. Appel [21], that Ising anisotropy was most favourable for ferromagnetic superconductors because transverse fluctuations would be pair-breaking as they would force scattering from one Fermi sheet to the opposite polarization Fermi sheet. However, this paper was written before the discovery of superconducting MgB_2 and the following boost of work on multigap superconductivity: we understand now that these transverse fluctuations do also induce exchange of Cooper pairs from one Fermi sheet to the other, which is favourable to superconductivity. And so the prediction is that 2D anisotropy (rather than uniaxial anisotropy) is the most favourable for ferromagnetic superconductors (maximizing both χ_{zz} and $\chi_{xx} - \chi_{yy}$).

The second reason is that, experimentally, all the systems where ferromagnetic superconductivity has been discovered did show a strong uniaxial anisotropy, confirming the prediction from [21]. But making statistics on few elements is always dangerous. We also found that reducing this uniaxial anisotropy in URhGe, using stress along the b-axis which increases the χ_{bb} susceptibility without changing χ_{cc} (for stress below 0.6 GPa), does increase T_{SC} in URhGe: a factor 2 between 0 and 1 GPa [22]. Ising anisotropy is probably not the most favourable, and larger T_{SC} ferromagnetic superconductors might be awaiting to be discovered.

Coming back to the consequences of (6.58) in terms of order parameter, with decoupling of the equations for η_x and η_y, the order parameter should have a line of node, (k_x or k_y or $k_z = 0$), a finite $\langle \mathbf{S} \rangle$ of order (at given $\mathbf{k}$)

$$\langle \mathbf{S}(\mathbf{k}) \rangle = \hbar \left(\frac{\delta}{\eta} \right) \left(\frac{k_i^2}{\langle k_i^2 \rangle} \right) \mathbf{e}_z ,$$

$$\delta = \frac{\eta_i^\uparrow - \eta_i^\downarrow}{2} ; \qquad \eta = \frac{\eta_i^\uparrow + \eta_i^\downarrow}{2} . \tag{6.60}$$

But it is not chiral ($\langle \mathbf{L} \rangle = 0$). The A state could be chiral if η_x and η_y remain coupled (in this model, if $\chi_{xy} \neq 0$), with

$$\langle \mathbf{L} \rangle = \hbar \frac{(\delta_y \eta_x + \delta_x \eta_y)\langle k_x^2 + k_y^2 \rangle}{(\eta_x^2 + \delta_x^2)\langle k_x^2 \rangle + (\eta_y^2 + \delta_y^2)\langle k_y^2 \rangle} \mathbf{e}_z . \tag{6.61}$$

Note that in principle too, if such is the case, $\mathbf{d}$ is probably not an ESP state any more, meaning that the Δ_0 component should be non-negligible and all expressions should be much more complex.

6.10 UTe$_2$

The last discovered p-wave superconductor is again an uranium-based system, also orthorhombic, and again close to a ferromagnetic instability but not ferromagnetic: UTe$_2$. This time, the T_{SC} is even more accessible: between 1.4 and 1.6 K from bulk measurements depending on the samples [23, 24]! Moreover, it presents similar astonishing field-reinforced superconductivity [23, 25, 26], with an absolute record (for such a low-T_{SC} system) of an upper critical field higher than 60 T [26]! An interesting point concerning the possible **d**-vector for such a paramagnetic system is the observation, on all samples, of a finite residual term of about half the normal state value of the specific heat C/T. The origin of this term is still unsettled, but an interesting proposal was that it would arise from a state similar to the A_1 state of superfluid ^{3}He (see Sect. 6.8.1.3).

Naturally, this can only happen if spin–orbit coupling is weak enough (otherwise, it would not be possible to form Cooper pairs on one Fermi surface and not on the other), and in this case, indeed, group theory classification leads to the possibility of such states (see Table 1 in [27])

$$\mathbf{d}(\mathbf{k}) = (1, i, 0)\varphi(\mathbf{k}) . \tag{6.62}$$

Moreover, because this is for a weak spin–orbit case, any rotation of the **d**-vector is a possible order parameter. According to (6.17), this means that the order parameter would be simply $\Delta(\mathbf{k}) = \Delta^{\uparrow}\varphi(\mathbf{k})$, with no other component. In the case of UTe$_2$, the largest susceptibility axis is the a-axis, so the quantization axis should be along **a**. In such a case, the Fermi surface with down-spin would remain unpaired, explaining the residual specific heat term.

Another consequence of such an order parameter is that it is non-unitary, with a finite spin for the Cooper pairs (along the a-axis). For the total system, this spin of the Cooper pairs would be compensated by that of the unpaired electrons (from the down-spin Fermi surface). This might lead to a total null magnetization. Nevertheless, in such a case, there is no reason that the state with such a **d**-vector would be stabilized, as half the condensation energy is lost compared to any other state, where pairing occurs on Fermi sheets of each spin direction. For such a state to be favoured, one needs some advantage of having this spin polarization in the superconducting state, so, for example, a form of coupling between the spin Cooper pairs and the normal state magnetization [27]. Then, this would also induce, like for superfluid ^{3}He in the A_1 state [28], a finite magnetization when entering the superconducting phase; globally, it is as if the system would become ferromagnetic on entering the superconducting state, with a (weak?) magnetization increasing linearly with temperature below T_{SC} [27]. Up to now, this has not been detected, and it remains to be settled if this (rather improbable) hypothesis is valid or not.

In any case, this last system beautifully confirms that p-wave superconductors are an incredible playground, where almost every new system brings its own share of surprise and stimulating challenges.

6.11   Proofs and Exercise Solutions

6.11.1   Proof of the Cayley–Klein Relation

Proof

$$\mathcal{R}_{\boldsymbol{\Omega}}(\mathbf{a}\cdot\boldsymbol{\sigma})\mathcal{R}_{-\boldsymbol{\Omega}}$$

$$= \left(\cos\Omega/2\,\mathbb{1} - i\,\sin\Omega/2\,\hat{\boldsymbol{\Omega}}\cdot\boldsymbol{\sigma}\right)(\mathbf{a}\cdot\boldsymbol{\sigma})\left(\cos\Omega/2\,\mathbb{1} + i\,\sin\Omega/2\,\hat{\boldsymbol{\Omega}}\cdot\boldsymbol{\sigma}\right)$$

$$= \left(\cos\Omega/2\,\mathbb{1} - i\,\sin\Omega/2\,\hat{\boldsymbol{\Omega}}\cdot\boldsymbol{\sigma}\right)\Big[\cos\Omega/2\,(\mathbf{a}\cdot\boldsymbol{\sigma})$$

$$+ i\,\sin\Omega/2\,\left((\mathbf{a}\cdot\hat{\boldsymbol{\Omega}})\mathbb{1} + i(\mathbf{a}\wedge\hat{\boldsymbol{\Omega}})\cdot\boldsymbol{\sigma}\right)\Big]$$

$$= \cos^2\Omega/2\,(\mathbf{a}\cdot\boldsymbol{\sigma}) + i\,\cos\Omega/2\,\sin\Omega/2\,(\mathbf{a}\cdot\hat{\boldsymbol{\Omega}})\mathbb{1}$$

$$- \cos\Omega/2\,\sin\Omega/2\,(\mathbf{a}\wedge\hat{\boldsymbol{\Omega}})\cdot\boldsymbol{\sigma}$$

$$- i\,\cos\Omega/2\,\sin\Omega/2\,\left(\mathbf{a}\cdot\hat{\boldsymbol{\Omega}}\mathbb{1} - i(\mathbf{a}\wedge\hat{\boldsymbol{\Omega}})\cdot\boldsymbol{\sigma}\right)$$

$$+ \sin^2\Omega/2\,\left((\mathbf{a}\cdot\hat{\boldsymbol{\Omega}})(\hat{\boldsymbol{\Omega}}\cdot\boldsymbol{\sigma}) + i^2\left(\hat{\boldsymbol{\Omega}}\wedge(\mathbf{a}\wedge\hat{\boldsymbol{\Omega}})\right)\cdot\boldsymbol{\sigma}\right)$$

$$= (\mathbf{a}\cdot\hat{\boldsymbol{\Omega}})(\hat{\boldsymbol{\Omega}}\cdot\boldsymbol{\sigma}) + \cos^2\Omega/2\left(\mathbf{a} - (\mathbf{a}\cdot\hat{\boldsymbol{\Omega}})\hat{\boldsymbol{\Omega}}\right)\cdot\boldsymbol{\sigma} + \sin\Omega\,(\mathbf{a}\wedge\hat{\boldsymbol{\Omega}})\cdot\boldsymbol{\sigma}$$

$$- \sin^2\Omega/2\left(\hat{\boldsymbol{\Omega}}\wedge(\mathbf{a}\wedge\hat{\boldsymbol{\Omega}})\right)\cdot\boldsymbol{\sigma}$$

$$= (\mathbf{a}\cdot\hat{\boldsymbol{\Omega}})(\hat{\boldsymbol{\Omega}}\cdot\boldsymbol{\sigma}) + \cos\Omega\left(\mathbf{a} - (\mathbf{a}\cdot\hat{\boldsymbol{\Omega}})\hat{\boldsymbol{\Omega}}\right)\cdot\boldsymbol{\sigma} + \sin\Omega\,(\mathbf{a}\wedge\hat{\boldsymbol{\Omega}})\cdot\boldsymbol{\sigma}$$

$$= \mathcal{R}_{\boldsymbol{\Omega}}(\mathbf{a})\cdot\boldsymbol{\sigma}.$$

6.11.2   Conservation of the Scalar Product under Rotation with the Definition (6.11)

Solution 6.1 Let us note that we can write, for any (complex) vectors $\mathbf{u}, \mathbf{a}, \mathbf{b}, \mathbf{c}, \mathbf{d}$,
...

$$u = (\mathbf{u}\cdot\hat{\boldsymbol{\Omega}})\boldsymbol{\Omega} + \left(\mathbf{u} - (\mathbf{u}\cdot\hat{\boldsymbol{\Omega}})\boldsymbol{\Omega}\right) = \mathbf{u}\mathbf{d}_{\parallel} + \mathbf{u}\mathbf{d}_{\perp}$$

$$\mathcal{R}(\mathbf{u}) = \mathbf{u}\mathbf{d}_{\parallel} + \mathcal{R}(\mathbf{u}\mathbf{d}_{\perp}) = \mathbf{u}\mathbf{d}_{\parallel} + \cos\Omega\,\mathbf{u}\mathbf{d}_{\perp} + \sin\Omega\,(\hat{\boldsymbol{\Omega}}\wedge\mathbf{u})$$

$$\mathbf{a}\wedge(\mathbf{b}\wedge\mathbf{c}) = \epsilon^{ijk}\epsilon^{klm}a_j b_l c_m = (\mathbf{a}\cdot\mathbf{c})\mathbf{b} - (\mathbf{a}\cdot\mathbf{b})\mathbf{c}$$

$$(\mathbf{a}\wedge\mathbf{b})\wedge\mathbf{c} = (\mathbf{a}\cdot\mathbf{c})\mathbf{b} - (\mathbf{c}\cdot\mathbf{b})\mathbf{a}.$$

$$\mathcal{R}(\mathbf{d})\cdot\mathcal{R}(\mathbf{u}) = \mathbf{d}_{\parallel}\cdot\mathbf{u}\mathbf{d}_{\parallel} + \mathcal{R}(\mathbf{d}_{\perp})\cdot\mathcal{R}(\mathbf{u}_{\perp})\quad\text{so}$$

$$\mathcal{R}(\mathbf{d}) \cdot \mathcal{R}(\mathbf{u})$$

$$= (\mathbf{d} \cdot \hat{\mathbf{\Omega}})(\mathbf{u} \cdot \hat{\mathbf{\Omega}})$$

$$+ \cos^2 \Omega \left[\mathbf{d}\cdot\mathbf{u} - (\mathbf{d} \cdot \hat{\mathbf{\Omega}})(\mathbf{u} \cdot \hat{\mathbf{\Omega}}) - \cancel{(\mathbf{u} \cdot \hat{\mathbf{\Omega}})(\mathbf{d} \cdot \hat{\mathbf{\Omega}})} + \cancel{(\mathbf{d} \cdot \hat{\mathbf{\Omega}})(\mathbf{u} \cdot \hat{\mathbf{\Omega}})} \right]$$

$$+ \sin^2 \Omega \left[\hat{\mathbf{\Omega}} \cdot (\mathbf{d} \wedge (\hat{\mathbf{\Omega}} \wedge u)) \right]$$

$$+ \sin \Omega \cos \Omega \left[\cancel{\mathbf{d} \cdot (\hat{\mathbf{\Omega}} \wedge \mathbf{u})} + \cancel{\mathbf{u} \cdot (\hat{\mathbf{\Omega}} \wedge \mathbf{d})} \right]$$

$$= (\mathbf{d} \cdot \hat{\mathbf{\Omega}})(\mathbf{u} \cdot \hat{\mathbf{\Omega}})$$

$$+ \cos^2 \Omega \left[\mathbf{d}\cdot\mathbf{u} - (\mathbf{d} \cdot \hat{\mathbf{\Omega}})(\mathbf{u} \cdot \hat{\mathbf{\Omega}}) \right]$$

$$+ \sin^2 \Omega \left[\mathbf{d}\cdot\mathbf{u} - (\mathbf{d} \cdot \hat{\mathbf{\Omega}}(\mathbf{u} \cdot \hat{\mathbf{\Omega}}) \right]$$

$$\mathcal{R}(\mathbf{d}) \cdot \mathcal{R}(\mathbf{u}) = \mathbf{d}\cdot\mathbf{u}.$$

6.11.3 Conservation of the Cross Product under Rotation with the Definition (6.11)

We want to check if $\mathcal{R}(\mathbf{u} \wedge \mathbf{d}) = \mathcal{R}(\mathbf{u}) \wedge \mathcal{R}(\mathbf{d})$. To evaluate $\mathcal{R}(\mathbf{u} \wedge \mathbf{d})$, let us decompose $\mathbf{u} \wedge \mathbf{d}$ in parallel and perpendicular parts to $\mathbf{\Omega}$ (called $(\mathbf{ud})_\parallel$ and $(\mathbf{ud})_\perp$)

$$\mathbf{u} \wedge \mathbf{d} = \underbrace{\cancel{((\mathbf{ud})_\parallel \wedge \mathbf{d}_\parallel)} + ((\mathbf{ud})_\perp \wedge \mathbf{d}_\perp)}_{(\mathbf{ud})_\parallel} + \underbrace{((\mathbf{ud})_\parallel \wedge \mathbf{d}_\perp) - (\mathbf{d}_\parallel \wedge (\mathbf{ud})_\perp)}_{(\mathbf{ud})_\perp}$$

$$= \underbrace{[(\mathbf{u} \wedge \mathbf{d}) \cdot \hat{\mathbf{\Omega}}]\hat{\mathbf{\Omega}}}_{(\mathbf{ud})_\parallel} + \underbrace{\left[(\mathbf{u} \cdot \hat{\mathbf{\Omega}})\hat{\mathbf{\Omega}} \wedge \mathbf{d}_\perp \right] - \left[(\mathbf{d} \cdot \hat{\mathbf{\Omega}})\hat{\mathbf{\Omega}} \wedge \mathbf{u}_\perp \right]}_{(\mathbf{ud})_\perp}$$

$$\mathcal{R}(\mathbf{u} \wedge \mathbf{d}) = (\mathbf{ud})_\parallel + \mathcal{R}\left((\mathbf{ud})_\perp\right)$$

$$\mathcal{R}((\mathbf{ud})_\perp)$$

$$= \cos \Omega \, (\mathbf{ud})_\perp + \sin \Omega \, \hat{\mathbf{\Omega}} \wedge (\mathbf{ud})_\perp$$

$$= (\mathbf{u} \cdot \hat{\mathbf{\Omega}})\hat{\mathbf{\Omega}} \wedge \left[\cos \Omega \, \mathbf{d} + \sin \Omega \, \hat{\mathbf{\Omega}} \wedge \mathbf{d} \right] - (\mathbf{d} \cdot \hat{\mathbf{\Omega}})\hat{\mathbf{\Omega}} \wedge \left[\cos \Omega \, \mathbf{u} + \sin \Omega \, \hat{\mathbf{\Omega}} \wedge \mathbf{u} \right]$$

$$= (\mathbf{u} \cdot \hat{\mathbf{\Omega}})\hat{\mathbf{\Omega}} \wedge \mathcal{R}(\mathbf{d}) - (\mathbf{d} \cdot \hat{\mathbf{\Omega}})\hat{\mathbf{\Omega}} \wedge \mathcal{R}(\mathbf{u})$$

$$= (\mathcal{R}(\mathbf{u}) \cdot \hat{\mathbf{\Omega}})\hat{\mathbf{\Omega}} \wedge \mathcal{R}(\mathbf{d}) - (\mathcal{R}(\mathbf{d}) \cdot \hat{\mathbf{\Omega}})\hat{\mathbf{\Omega}} \wedge \mathcal{R}(\mathbf{u})$$

$$= \hat{\mathbf{\Omega}} \wedge \left[\hat{\mathbf{\Omega}} \wedge (\mathcal{R}(\mathbf{d}) \wedge \mathcal{R}(\mathbf{u})) \right]$$

$$\mathcal{R}(\mathbf{u} \wedge \mathbf{d}) = \mathcal{R}(\mathbf{u}) \wedge \mathcal{R}(\mathbf{d}) + (\mathbf{ud})_\parallel - (\mathcal{R}(\mathbf{u}) \wedge \mathcal{R}(\mathbf{d}))_\parallel$$

$$(\mathcal{R}(\mathbf{u}) \wedge \mathcal{R}(\mathbf{d}))_\|$$
$$= \mathcal{R}(\mathbf{u})_\perp \wedge \mathcal{R}(\mathbf{d})_\perp$$
$$= \left(\cos \Omega\, \mathbf{u}_\perp + \sin \Omega\, (\hat{\boldsymbol{\Omega}} \wedge \mathbf{u}_\perp)\right) \wedge \left(\cos \Omega\, \mathbf{d}_\perp + \sin \Omega\, (\hat{\boldsymbol{\Omega}} \wedge \mathbf{d}_\perp)\right)$$
$$= \cos^2 \Omega\, (\mathbf{u}_\perp \wedge \mathbf{d}_\perp)$$
$$\quad + \sin^2 \Omega\, (\hat{\boldsymbol{\Omega}} \wedge \mathbf{u}_\perp) \wedge (\hat{\boldsymbol{\Omega}} \wedge \mathbf{d}_\perp)$$
$$\quad + \sin \Omega \cos \Omega \left(\mathbf{u}_\perp \wedge (\hat{\boldsymbol{\Omega}} \wedge \mathbf{d}_\perp) - \mathbf{d}_\perp \wedge (\hat{\boldsymbol{\Omega}} \wedge \mathbf{u}_\perp)\right)$$
$$= \cos^2 \Omega\, (\mathbf{ud})_\|$$
$$\quad + \sin^2 \Omega \left((\hat{\boldsymbol{\Omega}} \wedge \mathbf{u}_\perp)\cdot\mathbf{d}_\perp\right) \hat{\boldsymbol{\Omega}}$$
$$\quad + \sin \Omega \cos \Omega \left(\cancel{(\mathbf{u}_\perp\cdot\mathbf{d}_\perp)}\hat{\boldsymbol{\Omega}} - \cancel{(\mathbf{d}_\perp\cdot\mathbf{u}_\perp)}\hat{\boldsymbol{\Omega}}\right)$$
$$= \cos^2 \Omega\, (\mathbf{ud})_\| + \sin^2 \Omega \left((\mathbf{u} \wedge \mathbf{d}) \cdot \hat{\boldsymbol{\Omega}}\right) \hat{\boldsymbol{\Omega}} = (\mathbf{ud})_\|$$

So indeed, $\mathcal{R}(\mathbf{u} \wedge \mathbf{d}) = \mathcal{R}(\mathbf{u}) \wedge \mathcal{R}(\mathbf{d})$.

6.11.4   Rotation of the d-Vector of a Simple "Up-Up" State

Solution 6.2   For such a state, from (6.17), we get that

$$\mathbf{d} = \frac{1}{\psi} \begin{pmatrix} -\frac{1}{2}\Delta_\uparrow \\ -\frac{i}{2}\Delta_\uparrow \\ 0 \end{pmatrix}.$$

Changing the orientation of the (z) quantization axis amounts to a rotation of 6.17 by π around $\mathbf{e}_x$. From (6.11), we get

$$\mathcal{R}(\mathbf{d}) = d_x\, \mathbf{e}_x - (\mathbf{d} - d_x\, \mathbf{e}_x)$$
$$= -\mathbf{d},$$

which is indeed what we would expect!

6.11.5   Equivalence of ESP Unitary States and Pure $|S_z = 0\rangle$ States

Solution 6.3   Such an ESP state would have only $\Delta_\uparrow$ and $\Delta_\downarrow$ components, equal within a phase factor. Let us write $\Delta_\downarrow = e^{-2i\varphi}\Delta_\uparrow$. The **d**-vector of such a state will be [see (6.17)]

$$\mathbf{d} = \frac{1}{\psi}\begin{pmatrix} \frac{\Delta_\uparrow}{2}\left(e^{-2i\varphi}-1\right) \\ -\frac{i\Delta_\uparrow}{2}\left(1+e^{-2i\varphi}\right) \\ 0 \end{pmatrix} = \frac{1}{\psi}\begin{pmatrix} -ie^{-i\varphi}\Delta_\uparrow \sin\varphi \\ -ie^{-i\varphi}\Delta_\uparrow \cos\varphi \\ 0 \end{pmatrix}.$$

If we rotate this state by $\pi/2$ around a vector $\hat{\boldsymbol{\Omega}} = \begin{pmatrix} \cos\theta \\ \sin\theta \\ 0 \end{pmatrix}$ in the (x, y)-plane, we obtain from (6.11)

$$\mathcal{R}_\Omega(\mathbf{d}) = (\mathbf{d}\cdot\hat{\boldsymbol{\Omega}})\hat{\boldsymbol{\Omega}} + (\hat{\boldsymbol{\Omega}}\wedge\mathbf{d})$$
$$= \frac{1}{\psi}\begin{pmatrix} \sin(\theta+\varphi)e^{-i\varphi}\Delta_\uparrow\cos\theta \\ \sin(\theta+\varphi)e^{-i\varphi}\Delta_\uparrow\sin\theta \\ -ie^{-i\varphi}\Delta_\uparrow\cos(\theta+\varphi) \end{pmatrix}.$$

So indeed, if we choose $\theta = -\varphi$, we recover a pure $|S_z = 0\rangle$ state. Note that it is the phase between the $\Delta_\uparrow$ and $\Delta_\downarrow$ components of the order parameter, which determines the precise direction of the required $\pi/2$ rotation.

References

1. A.J. Leggett, A theoretical description of the new phases of liquid ^{3}He. Rev. Mod. Phys. **47**, 331 (1975). https://doi.org/10.1103/RevModPhys.47.331
2. V. Mineev, K. Samokhin, *Introduction to Unconventional Superconductivity* (Gordon and Breach Science Publishers, Amsterdam, 1999)
3. N. Mermin, V. Ambegaokar, The order parameter in liquid ^{3}He, in *Collective Properties of Physical Systems* (*Proceedings of the 24th Nobel Symposium*, Lerum 1973), ed. by B. Lundqvist, S. Lundqvist, and Runnström-Reio (Academic, New York, 1974), p. 97. https://doi.org/10.1016/B978-0-12-460350-9.50019-9
4. C. Kallin, J. Berlinsky, Chiral superconductors. Rep. Prog. Phys. **79**, 054502 (2016). https://doi.org/10.1088/0034-4885/79/5/054502
5. G.E. Volovik, V.P. Mineev, Investigation of singularities in superfluid He^3 in liquid crystals by the homotopic topology methods, Zh. Eksp. Teor. Fiz. **72**, 2256 (1977). [Sov. Phys. JETP **45**, 1186 (1977).]
6. R. Joynt, L. Taillefer, The superconducting phases of UPt$_3$. Rev. Mod. Phys. **74**, 235 (2002). https://doi.org/10.1103/RevModPhys.74.235
7. J. Sauls, The order parameter for the superconducting phases of UPt$_3$. Adv. Phys. **43**, 113 (1994). https://doi.org/10.1080/00018739400101475
8. E.R. Schemm, W.J. Gannon, C.M. Wishne, W.P. Halperin, A. Kapitulnik, Observation of broken time-reversal symmetry in the heavy-fermion superconductor UPt$_3$. Science **345**, 190 (2014). https://doi.org/10.1126/science.1248552
9. B.S. Shivaram, T.F. Rosenbaum, D.G. Hinks, Unusual angular and temperature dependence of the upper critical field in UPt$_3$. Phys. Rev. Lett. **57**, 1259 (1986). https://doi.org/10.1103/PhysRevLett.57.1259
10. Y. Machida, A. Itoh, Y. So, K. Izawa, Y. Haga, E. Yamamoto, N. Kimura, Y. Onuki, Y. Tsutsumi, K. Machida, Twofold spontaneous symmetry breaking in the heavy-fermion superconductor UPt$_3$. Phys. Rev. Lett. **108**, 157002 (2012). https://doi.org/10.1103/PhysRevLett.108.157002

11. H. Tou, Y. Kitaoka, K. Ishida, K. Asayama, N. Kimura, Y. Ōnuki, E. Yamamoto, Y. Haga, K. Maezawa, Nonunitary spin-triplet superconductivity in UPt_3: evidence from ^{195}Pt Knight shift study. Phys. Rev. Lett. **80**, 3129 (1998). https://doi.org/10.1103/PhysRevLett.80.3129

12. A.P. Mackenzie, Y. Maeno, The superconductivity of Sr_2RuO_4 and the physics of spin-triplet pairing. Rev. Mod. Phys. **75**, 657 (2003). https://doi.org/10.1103/RevModPhys.75.657

13. A. Pustogow, Y. Luo, A. Chronister, Y.S. Su, D.A. Sokolov, F. Jerzembeck, A.P. Mackenzie, C.W. Hicks, N. Kikugawa, S. Raghu, E.D. Bauer, S.E. Brown, Constraints on the superconducting order parameter in Sr_2RuO_4 from oxygen-17 nuclear magnetic resonance. Nature **574**, 72 (2019). https://doi.org/10.1038/s41586-019-1596-2

14. K. Ishida, M. Manago, Y. Maeno, Reduction of the ^{17}O knight shift in the superconducting state and the heat-up effect by NMR pulses on Sr_2RuO_4. J. Phys. Soc. Jpn. **89**, 034712 (2020). https://doi.org/10.7566/JPSJ.89.034712

15. S. Saxena, P. Agarwal, K. Ahilan, F. Grosche, R. Haselwimmer, M. Steiner, E. Pugh, I. Walker, S. Julian, P. Monthoux, G. Lonzarich, A. Huxley, I. Sheikin, D. Braithwaite, J. Flouquet, Superconductivity on the border of itinerant-electron ferromagnetism in UGe_2. Nature **406**, 587 (2000). https://doi.org/10.1038/35020500

16. D. Aoki, A. Huxley, E. Ressouche, D. Braithwaite, J. Flouquet, J.-P. Brison, E. Lhotel, C. Paulsen, Coexistence of superconductivity and ferromagnetism in URhGe. Nature **413**, 613 (2001). https://doi.org/10.1038/35098048

17. N.T. Huy, A. Gasparini, D.E. de Nijs, Y. Huang, J.C.P. Klaasse, T. Gortenmulder, A. de Visser, A. Hamann, T. Goerlach, H. von Loehneysen, Superconductivity on the border of weak itinerant ferromagnetism in UCoGe. Phys. Rev. Lett. **99**, 067006 (2007). https://doi.org/10.1103/PhysRevLett.99.067006

18. D. Aoki, K. Ishida, J. Flouquet, Review of U-based ferromagnetic superconductors: comparison between UGe_2, URhGe, and UCoGe. J. Phys. Soc. Jpn. **88**, 022001 (2019). https://doi.org/10.7566/JPSJ.88.022001

19. V.P. Mineev, Superconductivity in uranium ferromagnets. Phys. Usp. **60**, 121 (2017). https://doi.org/10.3367/ufne.2016.04.037771

20. P.G. de Gennes, *Superconductivity of Metals and Alloys* (Perseus Books, Reading, MA, new ed. of 2nd rev., 1999)

21. D. Fay, J. Appel, Coexistence of *p*-state superconductivity and itinerant ferromagnetism. Phys. Rev. B **22**, 3173 (1980). https://doi.org/10.1103/PhysRevB.22.3173

22. D. Braithwaite, D. Aoki, J.-P. Brison, J. Flouquet, G. Knebel, A. Nakamura, A. Pourret, Dimensionality driven enhancement of ferromagnetic superconductivity in URhGe. Phys. Rev. Lett. **120**, 037001 (2018). https://doi.org/10.1103/PhysRevLett.120.037001

23. S. Ran, C. Eckberg, Q.-P. Ding, Y. Furukawa, T. Metz, S.R. Saha, I.-L. Liu, M. Zic, H. Kim, J. Paglione, N.P. Butch, Nearly ferromagnetic spin-triplet superconductivity. Science **365**, 684 (2019). https://doi.org/10.1126/science.aav8645

24. D. Aoki, A. Nakamura, F. Honda, D. Li, Y. Homma, Y. Shimizu, Y.J. Sato, G. Knebel, J.-P. Brison, A. Pourret, D. Braithwaite, G. Lapertot, Q. Niu, M. Vališka, H. Harima, J. Flouquet, Unconventional superconductivity in heavy fermion UTe_2. J. Phys. Soc. Jpn. **88**, 043702 (2019). https://doi.org/10.7566/JPSJ.88.043702

25. G. Knebel, W. Knafo, A. Pourret, Q. Niu, M. Vališka, D. Braithwaite, G. Lapertot, M. Nardone, A. Zitouni, S. Mishra, I. Sheikin, G. Seyfarth, J.-P. Brison, D. Aoki, J. Flouquet, Field-reentrant superconductivity close to a metamagnetic transition in the heavy-fermion superconductor UTe_2. J. Phys. Soc. Jpn. **88**(6), 063707 (2019). https://doi.org/10.7566/JPSJ.88.063707

26. S. Ran, I.-L. Liu, Y.S. Eo, D.J. Campbell, P. Neves, W.T. Fuhrman, S.R. Saha, C. Eckberg, H. Kim, J. Paglione, D. Graf, J. Singleton, N.P. Butch, Extreme magnetic field-boosted superconductivity. Nat. Phys. **15**, 1250 (2019). https://doi.org/10.1038/s41567-019-0670-x

27. A.D. Hillier, J. Quintanilla, B. Mazidian, J.F. Annett, R. Cywinski, Nonunitary triplet pairing in the centrosymmetric superconductor $LaNiGa_2$. Phys. Rev. Lett. **109**, 097001 (2012). https://doi.org/10.1103/PhysRevLett.109.097001

28. S. Takagi, Susceptibility discontinuity at the He^3-A-normal transition. Prog. Theor. Phys. **51**, 1998 (1974). https://doi.org/10.1143/PTP.51.1998

Permissions

The contributors of this book come from diverse backgrounds, making this book a truly international effort. We would like to thank all the contributing authors for lending their expertise to make the book truly unique. They have played a crucial role in the development of this book. Without their invaluable contributions this book wouldn't have been possible. They have made vital efforts to compile up to date information on the varied aspects of this subject to make this book a valuable addition to the collection of many professionals and students.

This book was conceptualized with the vision of imparting up-to-date and integrated information in this field. To ensure the same, a matchless editorial board was set up. Every individual on the board went through rigorous rounds of assessment to prove their worth. After which they invested a large part of their time researching and compiling the most relevant data for our readers.

The editorial board has been involved in producing this book since its inception. They have spent rigorous hours researching and exploring the diverse topics which have resulted in the successful publishing of this book. They have passed on their knowledge of decades through this book. To expedite this challenging task, the publisher supported the team at every step. A small team of assistant editors was also appointed to further simplify the editing procedure and attain best results for the readers.

Apart from the editorial board, the designing team has also invested a significant amount of their time in understanding the subject and creating the most relevant covers. They scrutinized every image to scout for the most suitable representation of the subject and create an appropriate cover for the book.

The publishing team has been an ardent support to the editorial, designing and production team. Their endless efforts to recruit the best for this project, has resulted in the accomplishment of this book. They are a veteran in the field of academics and their pool of knowledge is as vast as their experience in printing. Their expertise and guidance has proved useful at every step. Their uncompromising quality standards have made this book an exceptional effort. Their encouragement from time to time has been an inspiration for everyone.

The publisher and the editorial board hope that this book will prove to be a valuable piece of knowledge for students, practitioners and scholars across the globe.

Index